AIRCRAFT PAINTING AND FINISHING

By Neal Carlson and IAP, Inc.

Library of Congress Cataloging-in-publication number: **93-24822**

JS312638B

Table of Contents

Howard Thomas 2-96

Preface _____ v

Introduction _____ vii

Chapter I   Metal Aircraft Finishing _____ 1
             A Surface Preparation _____ 1
             B. Primers _____ 3
             C. Topcoat Systems _____ 6

Chapter II   Fabric Aircraft Finishing _____ 9
             A. Cotton and Linen _____ 9
             B. Polyester (Dacron®, Ceconite®, etc.) _____ 12
             C. Polyester Over Plywood _____ 13
             D. Fiberglass Cloth _____ 13
             E. Built-Up Fiberglass Structure _____ 14
             F. Radomes _____ 14
             G. Repair to Fabric Structures _____ 14
             H. Repair to Fabric-Covered Plywood Surfaces _____ 14

Chapter III Application Procedure _____ 17
             A. Spray Painting _____ 17
             B. Aircraft Painting Sequence _____ 21
             C. Painting Safety _____ 23

Chapter IV   Finishing Problems _____ 29
             A. Filiform Corrosion _____ 29
             B. Color Matching _____ 29
             C. Difficulties With Acrylics _____ 31
             D. Difficulties With Polyurethanes _____ 31
             E. Difficulties With Dope _____ 31

Chapter V   Special Finishes and Finishing Products _____ 35
             A. High Visibility Finishes _____ 35
             B. Wrinkle Finish _____ 35
             C. Flat Black Lacquer _____ 35
             D. Wing Walk Compound _____ 35
             E. Acid-Proof Paint _____ 35
             F. Float Bottom Compound _____ 36
             G. Fuel Tank Sealer _____ 36
             H. Seam Paste _____ 36
             I. High Temperature Finishes _____ 36
             J. Rot-Inhibiting Sealer _____ 37

K. Spar Varnish _____ 37

L. Tube Oil _____ 37

    M. Thinners and Reducers _____ 37

    N. Rejuvenator _____ 38

    O. Spot Putty and Sanding Surfacer _____ 38

Chapter VI   Finishing Equipment _____ 39

    A. Paint Storage _____ 39

    B. Spray Booth or Spray Area _____ 39

    C. Air Compressors, Storage, and Distribution Lines _____ 39

    D. Spray Equipment _____ 40

    E. Respirators and Masks _____ 4

    F. Measuring Equipment _____ 41

    G. Mixing Equipment _____ 43

Appendix A _____ 45

Appendix B _____ 47

AppendixC _____ 49

Appendix D _____ 51

Appendix E _____ 53

Glossary _____ 55

Answers to Study Questions _____ 59

Final Examination _____ 61

Answers to Final Examination _____ 65

Preface

This book on *Aircraft Painting and Finishing* is one of a series of specialized training manuals prepared for aviation maintenance personnel.

This series is part of a programmed learning course developed and produced by International Aviation Publishers (IAP), one of the largest suppliers of aviation maintenance training materials in the world. This program is part of a continuing effort to improve the quality of education for aviation mechanics throughout the world.

The purpose of each IAP training series is to provide basic information on the operation and principles of the various aircraft systems and their components.

Specific information on detailed operation procedures should be obtained from the manufacturer through his appropriate maintenance manuals, and followed in detail for the best results.

This particular manual on *Aircraft Painting and Finishing* includes a series of carefully prepared questions and answers to emphasize key elements of the study, and to encourage you to continually test yourself for accuracy and retention as you use this book. A multiple choice final examination is included to allow you to test your comprehension of the total material.

For best results, the visual and audio portion should be reviewed first, either in the classroom under the direction of an experienced instructor, or by individual study; then this material should be reinforced with that included in this text.

Acknowledgements

The validity of any program such as this is enhanced immeasurably by the cooperation shown IAP by recognized experts in the field, and by the willingness of the various manufacturers to share their literature and answer countless questions in the preparation of these programs.

Mr. Neal Carlson, co-author of this text, has been in the aircraft finishing business since the production days of the E-2 Cub, the great-grandpappy, many times over, of the entire Piper line.

We would like to mention, especially, our appreciation for help given us by:

Amchem Products, Inc.
Binks Manufacturing Company
The DeVilbiss Company
E.I. DuPont De Nemours & Company
Finch Paint and Chemical Company
Randolph Products Company
Razorback Fabric, Inc.
U.S. Paint Lacquer and Chemical Co.

If you have any questions or comments regarding this manual, or any of the many other textbooks offered by IAP simply contact: Sales Department, IAP Inc.; Mailing Address: P.O. Box 10000, Casper, WY 82602-1000; Shipping Address: 7383 6WN Road, Casper, WY 82604-1835; or call toll free:(800) 443-9250; International, call: (307) 266-3838.

INTRODUCTION

Aircraft finishes have come a long way since banana oil was used to shrink and seal the fabric on the early wood, wire, and rag flying machines. Fabric-covered aircraft, while far less popular than in the past, have progressed through the multi-coated, hand-rubbed finishes on the stagger-wing Beeches, Wacos, and Stinsons to the more utilitarian finish applied to the agriculture or utility aircraft.

When labor was less costly and the air less polluted, metal airplanes glistened in their skins, clad with bare, pure aluminum. Today, many of these airplanes are protected with a hard, glossy skin of polyurethane, enamel, or acrylic that gives them their slick and shiny appearance, while protecting them from the ravages of the environment.

Aircraft finishes are important, not only for the attractive appearance they give the airplane, but for the protection they afford the lightweight, highly reactive metals of which the structure is made.

When an airplane leaves the factory, it has been given a finish that is both decorative and protective. It is the responsibility of the maintenance personnel to see to it that this finish is maintained in such a way that it will keep its beauty and continue this protection. If the airplane is to be refinished, the A&P must properly prepare the surface and apply a new finish that will protect at least as well as the original.

CHAPTER I

Metal Aircraft Finishing

A. Surface Preparation

It goes without saying that no finish will last long if the metal surface has not been properly prepared. The metal must be thoroughly cleaned and microscopically roughened to provide a bond for the finish. A primer applied to the metal provides a sandwich to which the topcoats can adhere. If the metal has already been painted, this paint must be thoroughly reconditioned or completely removed before a new finishing system can be applied.

1. Paint Stripping

There are two types of paint strippers which can be used to remove the finish from an airplane. The solvent type, which is a clear liquid, is not very effective for stripping an airplane because of the fast rate of evaporation of its active solvents. These solvents do not have time to penetrate the film.

Wax-type removers are most generally used when stripping an entire airplane because the wax holds the active solvents against the surface until they penetrate it. Methylene chloride is the active agent in this type, and it penetrates the film of enamels or some primers to expand them so they pucker up and break their bond with the metal. After the bond has been broken, the wax gets between the film and the metal, preventing its resticking. Never remove, or attempt to remove, the stripper until all of the area has puckered up, or is completely softened. Flushing the stripper before it has finished its work defeats its effectiveness. If an area dries before it puckers or softens, apply some more stripper and allow it to remain until its action is complete.

To properly strip a surface, the remover is applied with a bristle brush, a non-atomizing spray, or a roller. If it is brushed on, a heavy, wet coat should be applied, brushing only in one direction. After this has been done, lay an inexpensive polyethylene drop cloth over the surface to hold the solvents until they have had ample time to penetrate the film.

Acrylic lacquer will not expand or wrinkle when the stripper works on it. It will only soften. As an area is softened, the drop cloth should be rolled back, exposing a small section of the softened finish. This is scraped off with a piece of Plexiglass™ or a rubber squeegee, and the cleaned area washed with a rag wet with methyl-ethyl-ketone (MEK) or acetone. Roll the drop cloth back and remove more finish.

One of the most important parts of the paint stripping process is the complete removal of every trace of wax left on the surface from the stripper. Careful scrubbing with acetone or MEK after removing the acrylics generally leaves the surface free from wax, provided all of the faying strips and the area around the rivets, fittings, and joints are flushed out with the solvents in a power spray gun.

Enamel or polyurethane residue must be flushed off with water, and the entire surface scrubbed with a good solvent. MEK or acetone is generally suitable, but a less expensive solvent such a toluol or xylol is more desirable. Lacquer thinner is not satisfactory because it will not absorb the wax; it will only spread it around. Any wax left on the surface will tend to be absorbed by the solvents in the finish and brought up into the system and locked in, preventing its drying.

Polyurethane film is readily attacked by the solvents, but if it is held against the surface long enough, the active agent will loosen the bond to the primer and release the film. The surface formed by a properly converted wash primer will not be damaged by the paint stripper; it can remain on the finish until the polyurethane has completely puckered up or lifted before flushing off the stripper and paint residue.

No prepared paint remover should be used on aircraft fabric or be allowed to come in contact with any fiberglass reinforced parts such as radomes, radio antenna, or any component such as fiberglass-reinforced wheel pants or wing tips. The active agents will attack and soften the binder in these parts.

CAUTION: Any time you use a paint stripper, always wear protective goggles and rubber gloves. If any stripper is splashed on your skin, wash it off immediately with water; and if any comes in contact with your eyes, flood them repeatedly with water and **CALL A PHYSICIAN.**

2. Corrosion Removal

Any trace of the white powder which indicates corrosion must be critically examined.

Corrosion of aluminum or magnesium is essentially an electro-chemical process in which the alloying agents in the metal have reacted with moisture and/or oxygen on the surface to form an electrical battery and generate a flow of electrons. The chemical action which caused these electrons to flow has converted some of the metal into a porous salt which has no physical strength. Corrosion, once it starts, will often continue until the skin or component is damaged beyond repair. The manual entitled *Aircraft Corrosion Control* deals in detail with this problem—its cause and correction. If corrosion is found on the airplane you have stripped, it is recommended that you consult this book.

If corrosion is found, every trace must be removed with fine sandpaper (no emery), aluminum wool, or a nylon scrubber.

CAUTION: Never use steel wool or a steel brush to remove corrosion from aluminum, as tiny bits of steel will embed in the aluminum and cause much worse corrosion than you had to begin with.

3. Conversion Coatings

After every trace of corrosion has been removed, it must be determined whether or not the structure has been damaged enough to require replacement or reinforcement. If the damage is relatively superficial, the metal may be treated with a conversion coating. This is essentially a phosphoric acid etchant which reacts with the metal to convert into a phosphate film over the metal and prevents recurrence of the corrosion.

The acid content of these materials is so low that a thorough flushing of the surface with water followed by air drying is sufficient to remove all traces of any unconverted acid. The extremely thin phosphate film left by this conversion provides a good bond for subsequent primers or topcoats.

Conversion coatings are applied to surfaces of new clad aluminum to microscopically roughen and prepare them so the additional coats will adhere.

4. Corrosion Protection For Dissimilar Metals

Any time aluminum and magnesium are to be joined, the magnesium should be treated with a chromic acid brush-on treatment similar to the Dow 19 treatment. Mix 1⅓ ounces of chromic anhydride (CrO_3) with one ounce of calcium sulfate ($CaSO_4$) in enough water to make one gallon. Brush this on the magnesium and let it stay for one to three minutes and rinse it off with cold water. After it is thoroughly dry, treat the surface with a wash primer, and after this has cured, coat it with an epoxy primer.

Use zinc chromate primer as a dielectric between the two metals. Spray both pieces and join them while the primer is wet. Wipe off the excess after the parts are together, and finish as required.

B. Primers

After the surface has been properly pre-treated, a primer is applied to provide a good bond between the metal and the topcoats. For years, zinc chromate has been the standard primer for aircraft use because of its good corrosion resistance. But, since it does not provide as good a bond to the surface as some of the new primers, its use is decreasing. Two-component epoxy primer is recommended.

1. Wash Primer

High-volume production of all-metal aircraft has brought about the development of a primer which provides a good bond between the metal and the finish, and which allows the topcoat to be applied after only about a half-hour cure. These primers can be used on aluminum, magnesium, steel or on fiberglass. Acrylic or enamel topcoats can be applied directly over the wash primer, but for maximum protection, such as required for seaplanes or agricultural aircraft, an epoxy primer should be applied over it.

When wash primer is applied over a properly cured conversion coating, the organic film of the wash primer bonds with the inorganic film and provides excellent adhesion between the topcoat and the surface. It also provides good protection for the metal.

Wash primer is a three-component material. Four parts of primer are mixed with one part of acid diluent and four parts of thinner and allowed to stand for twenty minutes to begin its curing action. Restir, and spray on the surface. The viscosity is adjusted by the addition of more thinner in order to get the extremely thin film required, but never use more than 8 parts of thinner to 4 parts of primer.

Wash primers should be applied with a film thickness of not more than 0.3 mil (0-0003 in., 0.0076 mm). This can be determined by looking at the surface. A film of proper thickness will not nearly hide the surface, but will give a slight amber cast to the aluminum.

One of the main reasons wash primers are so popular for production is the fact that they may be topcoated in a short time after their application without the finish sinking in and losing its gloss. The acid requires about thirty minutes to convert into the phosphate film, so the topcoat must not be applied until the conversion has finished; but it must be applied within eight hours, or the glaze on the primer will be so hard that the topcoat will not adhere to it. Every effort should be made to topcoat wash primer within this time frame of one to eight hours; but if it is absolutely impossible to finish within this period, another coat of primer must be applied over this first one. Omit the acid when mixing the primer for the second coat and when it is going to be used over fiberglass or plastic.

About the most critical aspect of the application of wash primers is the necessity of having sufficient moisture in the air to properly convert the acid into the phosphate film. It has been proven by much research that proper conversion requires nine-hundredths of a pound of water for every pound of dry air during the application of the primer. It is not really difficult to know exactly how much water is in the air if you refer to the chart in Figure 1.

This is a modification of a relative humidity chart, and in order to use it, you must have two mercury thermometers. Wrap a cotton wick around the bulb of one, and, with the thermometers placed side by side in the spray booth, blow air from the spray gun across them. This will evaporate the water from the wick and lower the temperature measured on that thermometer. Locate the temperature of the thermometer without the wick (dry-bulb) across the bottom of the chart, and follow this line up until it is crossed by the slanting line representing the temperature of the wet-bulb thermometer. Read the amount of water on the horizontal line through this intersecting point. Let's assume, for instance, that the dry-bulb temperature is 70° degrees F and the wet-bulb temperature is 60° degrees. These two lines cross above the horizontal line indicating 0.09 pound of water per pound of dry air, actually at about 0.095 pound. This means that there is enough water in the air to properly convert the acid in the primer.

If there is not enough water in the air for proper conversion, the finish will trap active acid against the metal. In order to prevent this, and the subsequent danger of corrosion, water may be added to the thinner. The thinner for wash primer is primarily an alcohol, and it will accept water. If there is

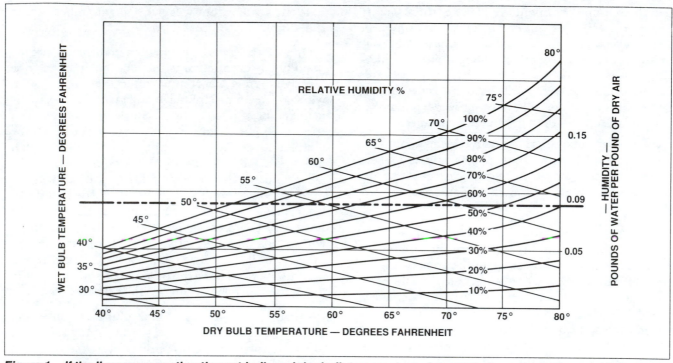

Figure 1. *If the lines representing the wet bulb and dry bulb temperatures intersect above the 0.09 pound line, there is sufficient water in the air. If the intersection is between the 0.09 and 0.05 pound lines, you may add one ounce of distilled water to one gallon of the thinner. If the intersection is below the 0.05 pound line, a maximum of two ounces of distilled water may be added to each gallon of thinner to aid the cure of the primer.*

somewhere between 0.05 and 0.09 pound of water to each pound of air, you may add one ounce of distilled water to each gallon of thinner. If there is less water than 0.05 pound per pound of dry air, you may add two ounces to each gallon of thinner; but this is the maximum amount permissible under any circumstances.

Acrylic lacquer applied over an improperly cured wash primer is porous enough to allow moisture from a heavy dew to penetrate the film and unite with the free acid and convert it. In the process of doing this, the paint, primer and all, will be lifted from the surface in the form of blisters. If these blisters appear shortly after the painting has been finished, allow the airplane to sit in the sunshine and get thoroughly warmed. These blisters will go down and the surface will be smooth again. The acid has now received sufficient water for its conversion and the primer will have its proper cure; the finish has not been damaged, provided the condition was not initially too severe.

Another condition could exist within the wash primer if it is not sufficiently converted and is covered with an epoxy or zinc chromate primer or a polyurethane topcoat. These finishes are not as porous as acrylics and will not allow sufficient water

to enter to complete the conversion, but will allow enough to penetrate to react with the acid and the metal to form filiform corrosion. This is simply corrosion having a thread-like form in which the acid and water have reacted with the metal and formed a

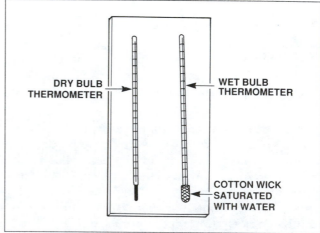

Figure 2. *Saturate the wick surrounding the wet bulb thermometer with water, and blow air across the two thermometers. The difference between the two readings is located on the table in Figure 1 to determine the amount of water in the air.*

4

salt, destroying the metal. Once filiform corrosion is detected under any of these films, all of the finish must be stripped and the entire surface treated to remove every trace of the corrosion.

2. Epoxy Primer

The most popular primer for use under the new polyurethane finishes and for any finish where the maximum corrosion protection is required is one of the epoxy primers. These are two component materials which produce a tough, dope proof sandwich coat between the finish and the surface. Epoxy primers may be used on aluminum, magnesium, steel or fiberglass, and for the maximum corrosion protection can be applied over wash primers.

Epoxy primers are not used for high-volume production aircraft because of the long time period required for them to develop their hold-out. This is the hardness required of a primer to prevent the topcoat sinking into it and distorting its surface so it loses its gloss. A waiting period of at least five hours, and preferably over night, is required before topcoating epoxy with acrylic or with enamel. These finishes will soften the primer and lose gloss if they are applied before the primer has had its full time to cure. Polyurethane enamels are compatible with the epoxy primers and will not re- lift them, so they may be applied after a wait of only about one hour. The polyurethane finish softens the surface of the epoxy and forms a chemical bond with the primer.

If you ever have to wait more than 24 hours between the time the primer is sprayed on the surface and the application of the topcoat, the epoxy will have to have its surface glaze broken by scuffing it with crumpled kraft paper, number 600 sandpaper, or a Scotch-Brite pad. This hard surface, if not roughed up, will not soften enough to allow the finish to adhere.

Wash the surface with an acrylic lacquer thinner or toluol. MEK is excellent, but it is more costly. Thoroughly mix one part of epoxy primer with one part of primer catalyst or mixing liquid, stirring the components separately, and then stirring them together. Add 1-1/2 parts of thinner and allow the mixture to age for twenty minutes. Restir and apply to the surface. This should be sprayed on with one light, even coat to give a film thickness of about half a mil (0.0005 in, or 0.013 mm), just thick enough to slightly color the metal. The catalyst for epoxy primer is quite reactive to moisture and the container should be kept tightly closed. If the lid should be left off of the catalyst for some time, and then the container resealed, the moisture which has been absorbed into the material can cause an action that could burst the can. Epoxy primers should be used within six hours after they are mixed. After the spraying operation is completed, the spray gun and hose must be cleaned out with the same thinner used for mixing, or with MEK.

3. Zinc Chromate Primer

MIL-P-8585 zinc chromate primer is just about one of the best known finishing materials used by A&P technicians. Its familiar green or yellow is "what airplane primer is supposed to look like." This has been the attitude for years; but it is losing ground now to the faster wash primers with their better adhesion, or to the far more durable epoxy primers. Zinc chromate is still a good primer as far as corrosion resistance is concerned, but it is inferior to the others with regard to adhesion. Where it is desired to use zinc chromate, it can be effectively sprayed over a surface which has been properly treated with a conversion coating such as Alodine. The Alodine provides for the adhesion, so the corrosion-inhibiting qualities of the zinc chromate can be used.

Zinc chromate is held in an alkyd resin. This does not produce an absolutely tight surface, but allows a small amount of water to enter the film and free some of the chromate ions, preventing, or at least inhibiting, the formation of corrosion on the surface it protects. It may be thinned for spraying with a proprietary thinner or with toluol.

Zinc chromate is available in both the familiar yellow and green colors. The primers are the same except for a touch of black pigment put into yellow primer to make it green. Red iron oxide may be added to zinc chromate primer to produce a hard, tough, protective film.

While zinc chromate has a wide usage as a primer for both aluminum and steel, it does have some limitations. It should not be applied over a wash primer, unless you are absolutely sure all of the phosphoric acid has been converted into the phosphate film. The zinc chromate primer will tend to entrap water and allow the formation of filiform corrosion. Zinc chromate should not be used as a base coat for acrylic lacquers, as the solvents in the acrylic will lift the zinc chromate unless it has aged for several days.

C. Topcoat Systems

1. Enamel

It can sometimes be a rather fine point whether a material is a lacquer or an enamel, but a pretty general definition identifies a lacquer as a finish which cures by the evaporation of its solvents, and one which can always be put back into its original condition by the use of thinners. An enamel cures by the conversion of some of its solvents, by heat, oxidation, or by catalytic action. One test is to rub a little of the thinner used to reduce the material for spraying over some of the dried material. If it softens the film, the material is a lacquer. If it does not, it is in all probability an enamel.

The older enamels were essentially pigments suspended in an oil-type varnish. These are no longer used in any production aircraft, but may be encountered when restoring old aircraft to their original finish. These enamels air-dry to touch by flashing off, or the evaporation of the solvents; the true cure is by oxidation or polymerization of the resins.

The only acrylic enamels which can be considered as suitable aircraft enamels are those whose cure is produced by baking; that is, by heat conversion.

Conventional enamels are supplied at a solids content of about 45 to 50 percent; acrylics have about 2/3 this amount of solids, considerably less than the polyurethanes, which have at least 60 percent solids.

Enamels are reduced or thinned with a proprietary enamel reducer or with toluol by a ratio from 20 parts of enamel to one part thinner, to a maximum of five parts enamel to one part thinner. Ten parts enamel to one part thinner is about the typical reducing ratio. The thinners used have a high solvency and are used to reduce the viscosity of the material, not to thin the solids. A five to one or a ten to one reduction allows you to spray on a thin coat of material with high solids content.

Wash primer or epoxy primer should be applied to the surface, allowed to dry, then have its glaze broken by scuffing it with crumpled kraft paper. Spray on a light mist coat of enamel, and allow the thinners to flash off, which takes about fifteen minutes. Follow with a full wet cross-coat and allow to dry for about forty-eight hours before taping or masking.

2. Acrylic Lacquers

High-volume production of aircraft has brought out a requirement for a finish that has best been met with acrylic lacquers. This material has a low solids content, compared with either conventional enamel or polyurethane, and it may be applied over wash primer, favored for new production, or over any of the epoxy primers. After the primer is thoroughly dry, rub it down with clean, dry kraft paper and apply the finish. A white base coat should be applied to assure the proper color match for the finish.

The low solids content of acrylics makes them somewhat tricky in their application. Their low viscosity and poor hiding qualities would seem to favor heavy coats, but not so; they should be thinned, using four parts of material to five parts thinner. This seems more thinner than actually required, but is good for a starting point. Adjust the amount of thinner to get the best coat. Spray on a very light tack coat, then follow with at least three cross-coats, allowing about a half hour drying time between coats. If the material is too heavy, pinholes or orange peel are likely to show up in the finish. The gloss in the final coat may be improved by adding about a fourth as much retarder as you have thinner in the material. If retarder has been used in the final coat, the finish should be allowed to dry overnight before taping or masking.

3. Polyurethane Enamel

One of the most durable and attractive finishes for modern, high-speed, high- altitude airplanes is the polyurethane enamel system. This hard, chemically resistant finish finds wide application with agricultural aircraft, seaplanes, and others which operate in hostile environments.

Polyurethane enamel is a two-part, chemically cured finish having a very high solids content, at least 60%. The high gloss inherent with this system is primarily due to the slow-flowing resins used. The thinners flash off quickly but the resins continue to flow for three to five days. It is this long flow- out time and the even cure throughout the film that give the pigment and the film time to form a truly flat surface, one that reflects light and has the glossy "wet" look which makes them so popular.

Polyurethane finish is used on agricultural aircraft and seaplanes because of its abrasion resistance and resistance to chemical attack. Skydrol™ hydraulic fluid, which quite actively attacks and softens other finishes, has only animal effect on polyurethanes. Even acetone will not dull the finish. Paint strippers must be held to the surface for a good while to give the active ingredients time to break through the film and attack the primer.

Wash primers may be used for polyurethanes, but for corrosion resistance, epoxy primers are recommended. The best undercoating is a conversion coating applied with care, using the manufacturer's recommendation *to the letter*, with an epoxy primer applied over it.

Polyurethane enamel is mixed with its catalyst in the proportion specified in the mixing instructions, usually in a one-to-one ratio. It is allowed to stand for about fifteen minutes as part of its induction period. In this time the curing action is started. The primary purpose of this waiting period is to aid in the inter-mixing or blending of the two components.

After this induction period, the material is stirred and mixed with reducer to the proper viscosity for spraying. This is measured with a number 2 Zahn cup and is between 18 and 20 seconds. There is more about this method of viscosity measurement in the section on paint shop equipment. When you have the proper viscosity, spray on a very light tack coat, lighter than with a conventional enamel. Allow it to set for about fifteen minutes so the thinner can flash off, or evaporate, and spray on a full wet cross-coat.

The main problem with the application of polyurethane lies in getting it on too thick. A film thickness of about 1.5 mils (one-and-a-half thousandths of an inch) is about maximum for all areas except for those subject to excessive erosion, such as leading edges. Too thick a film which might build up in the faying strips can crack because of the loss of flexibility. A good practical way to tell when you have enough material is to spray until you feel that one more pass will be just right, then quit right there, before you make that one more pass. The high solids content of polyurethane, its slow drying, and low surface tension allow the finish to crawl for an hour or so after it has been put on. If you can still see the metal when you think you have almost enough, don't worry; it will flow out and cover it. Almost no polyurethane job will look good until the next day, because it is still flowing. It will actually flow for about three to five days. It will be hard in this time, and the airplane may be flown in good weather, but the paint below the surface is still moving.

Masking tape may be applied after 12 hours under the most ideal conditions, but it is far better if you can wait 24 hours after application of the finish; it should be removed as soon after the trim is sprayed as possible. If it is left on the surface for a day or so, it will be almost impossible to remove.

Both polyurethane enamel and the epoxy primer that sandwiches the film to the surface are catalyzed materials. They should be mixed and used within six hours. If they are not applied within this time, they will not have the full gloss because of the reduced flow time. If it is impossible to spray all of the polyurethane within the six hour time period, careful addition of reducer can add a couple of hours to the useful life of the material.

The catalysts used for these primers and finishes are highly reactive to moisture, and the cans should be recapped immediately after using. If a can of the catalyst is allowed to remain open for a period of time, and is then resealed, the moisture in the can will activate it, and swell it up so much there is danger of the can bursting. High humidity and/or heat accelerate the cure.

All catalyzed material must be removed from the pressure pot, the hose, and the gun, immediately upon completion of the spraying operation, and the equipment thoroughly washed. If any of this material is allowed to remain overnight, it will solidify and ruin the equipment.

NOTE: Polyurethane may be injurious to your health. See Figure 25. Wear proper safety equipment and clothing. Also see the section on Polyurethane Paint Safety. This is in Chapter III in this book.

QUESTIONS

1. What is the purpose of the wax in a paint stripper?

2. How does paint stripper affect acrylic lacquer?

3. How does paint stripper remove a polyurethane finish?

4. What does a conversion coating do to the aluminum skin?

5. What is the reason for allowing wash primer to set for twenty minutes between mixing and spraying?

6. How thick should a coat of wash primer be?

7. How soon after spraying on wash primer can the finish coat be applied?

8. What should you do if you have not top coated wash primer within eight hours?

9. What should you do to be assured of enough water to properly convert the acid in wash primer if the dry-bulb temperature is 60° F and the wet-bulb temperature is 45° F?

10. How thick should an epoxy primer be on the surface of a metal?

11. What can be used to thin zinc chromate primer?

12. What will happen if acrylic lacquer is sprayed over freshly applied zinc chromate?

13. How much thinner is normally used to thin enamel for spraying?

14. How much thinner is used to thin acrylic lacquer for spraying?

15. Why does a polyurethane finish usually not look good immediately after you finish spraying it?

16. How long should you wait after putting on a polyurethane finish before it can be taped and masked?

CHAPTER II

Fabric Aircraft Finishing

Although almost all aircraft in current production are of all-metal construction, there are still some rag-wing airplanes being built, and there are definitely many of them still flying. Re-covering and refinishing these airplanes is still a bread and-butter job for A&P technicians and must be done right.

This book is on aircraft finishing, not re-covering. Any re-covering of certificated aircraft must be done according to the methods covered in AC 43.13-1A, Chapter 3. Now, we all know that AC 43.13A lags the technology of the industry, and no instructions are given relative to the newer fabrics such as polyester or pre-treated fiberglass cloth. For the application of these fabrics, you must use the instructions included in the Supplemental Type Certificate, under which the recovering of a particular airplane with these fabrics is approved.

We will consider the finish systems used for the four most popular fabrics: cotton, linen, polyester, and fiberglass.

A. Cotton And Linen

Grade A long-staple cotton meeting TSO-C-15 requirements is the standard fabric used for aircraft covering. Replacement of this fabric with more Grade A may be done, using the information in 43.13-1A and following the procedure of the aircraft manufacturer. A good Grade A or linen job should last as long as the structure below the fabric can afford to go without inspection. This is somewhere around five to eight years, depending on the protection the airplane is given. These fabrics can deteriorate in strength from Grade A's original 80 pounds per inch to 70% of this, or 56 pounds per inch, before it must be replaced. Of course, if the airplane only requires intermediate fabric, having a new fabric strength of 65 pounds per inch, and is covered with Grade A, it can deteriorate to 46 pounds before it must be replaced.

Linen meeting the British standards 7F1 is listed in 43.13-1A as an acceptable material for recovering airplanes requiring Grade A. It can replace Grade A without having to use an STC.

As in all aspects of aviation maintenance, it is only when the material actually becomes a part of a certificated airplane that the FAA shows an interest in it. For this reason, before re-covering an airplane, make a pull test of the fabric to be sure the new material has the required strength. Buy your material from a reputable dealer who sells only fresh fabric. Buy it only when the structure is ready to cover, so it will not age in your shop, and store it in a cool, dry, dark room until you are actually ready to put it on the airplane.

Dope has a tendency to turn acidic when it is old, and for this reason you should be sure the dope you buy is fresh. Old dope and reclaimed solvents have been sold by unscrupulous dealers through aviation classified ads and have ruined fabric jobs before the airplane has had a chance to fly. The difference between the cost of good, fresh, first quality supplies and those of unknown origin or quality is so small in comparison with the total re-covering cost that the latter should not be considered. It is the responsibility of the A&P technician to determine that the fabric and the dope will provide an airworthy cover job.

Most fixed base maintenance shops are not staffed or equipped for re-covering airplanes on a regular basis, and because of the relatively low return for the amount of labor involved, these jobs are often used as fill-ins during slack periods of the more lucrative work. One thing to consider when scheduling fabric replacement is to allow at least enough time to take the fabric up through the aluminum dope with no time breaks in the procedure. If the fabric is put on the structure and allowed to sit for even a few days with no dope on it, fungus spores can get into the fabric and cause deterioration in a short time.

1. Fabric Attachment

Before the fabric is attached to the structure, all of the necessary repairs must be made and inspected, and the structure primed. Metal tube structures, such as the fuselage and empennage, area after

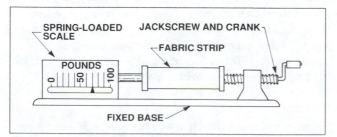

Figure 3. A pull test should be performed on all fabric before any of it is used on an airplane.

careful cleaning and sandblasting, primed with an epoxy primer. This hard, durable, dope-proof finish will protect the metal structure until it is next re-covered. Don't try to economize here by using a less durable primer, because rust in the tubular structure will surely shorten the life of the re-covered airplane. Wood structure is protected by spar varnish. This material hardens as it reacts with oxygen in the air and provides a good water-resistant surface that will not be lifted by the dope. White dope-proof paint was used in the past over any structure which would come in contact with the dope. This is simply a slow-drying white enamel and is not necessary if well-cured spar varnish is on the surface.

AC 43.13-1A considers fabric attachment primarily by sewing. While this has been the best way in the past and is the only way for some structures, heavy-bodied nitrate cement similar to nitrate dope, except for the special cotton used in its production, may be used to attach the fabric to the fuselage longerons, the empennage, and the wing trailing edges. (If there is any question regarding the legality of using a cemented seam instead of a sewn seam, be sure to check with your local FAA district office.)

Apply a heavy coat of this cement to the structural members to which the fabric will attach, and allow it to dry. Brush a coat of the cement on the underside of the fabric, and a fresh coat on the structure, and embed the fabric in the cement. Work all of the air bubbles out with your fingers and smooth the fabric to the surface. Be careful not to allow the cement to drip inside the fabric, or it will cause a rough spot which will show through the final finish. Apply this cement with a small brush. Either nitrate or butyrate dope thinner can be used to thin this cement or to clean the brush.

Fabric is attached to the structure in some production shops with a contact cement, but this is not recommended for use in the field because of some of its unique application problems.

2. Prime Coats

After the cotton fabric has been properly attached to the surface and shrunk with distilled or demineralized water to remove any wrinkles, the prime coats of dope are applied.

Butyrate dope is normally considered to be the best dope for finishing fabric because of its slower burning rate, but nitrate dope has better encapsulating properties and may be used to advantage for the first coat. The dope for this coat is thinned enough to allow good penetration without dripping through on the inside. This will be with about equal parts of thinner and clear dope.

Both Grade A cotton and linen are organic fabrics and are subject to mildew attack, so a fungicidal treatment should be given the dope used for the first coat. Before thinning the dope, mix four ounces of fungicidal paste with one gallon of dope. Mix the dope into the paste, not the paste into the dope. Be sure it dissolves completely and is thoroughly mixed before brushing it into the fabric. All fungicidal ingredients are poisonous, but since there is such a small amount in the dope, normal precautions should prevent any ill effects from using this material.

Mildew or mold forms on the inside of the fabric, so the fungicide-treated dope should penetrate the fabric completely. With this first coat, brush the dope into the fabric, whereas the other coats will be brushed onto the surface.

Allow the fungicide coat to dry for at least an hour; then, using butyrate dope, apply the surface tape, drainage grommets, and inspection rings. Lay the tape in a heavy coat of dope, and saturate it by brushing dope into its underside as it is laid. Work it down to the surface smoothly and rub out any air pockets with your fingers. Brush clear dope over the tape. Allow this to dry for at least an hour and brush a coat of clear butyrate dope over the entire surface. This, the second full coat, is brushed on, using the brush to transport the material to the surface. Long, smooth strokes should be used instead of working the brush back and forth.

Butyrate dope will penetrate the film of nitrate and bond to it, and since nitrate wraps around the fibers so much better than butyrate, using nitrate for the penetration coat and butyrate for all the other coats is the preferred finishing sequence. Once butyrate dope has been applied, nitrate cannot be used over it, as the solvents in the nitrate will not penetrate the butyrate film.

The dope for the second coat should be thinned just enough that it will not rope or pull.

3. Fill Coats

After the second coat of clear dope has completely dried, the fabric will be loose and baggy, and the nap will stand up. Very lightly sand this nap off with dry sandpaper.

CAUTION: Be sure the structure is electrically grounded to a cold water pipe or the hangar structure to prevent sparks from static electricity.

Spray on two full wet cross-coats of clear butyrate dope, thinned enough for proper spraying. This will be from about five parts of dope to one part thinner, to one part each, depending on the spray equipment you are using.

The ultraviolet rays from the sun damage the fibers of the cloth and the dope film, so a coat of aluminum dope must cover the undercoats of clear dope to provide protection. Aluminum dope is prepared by mixing one pound of aluminum paste with five gallons of clear dope before thinning. (Smaller quantities can be made by mixing 3-1/2 ounces by weight of aluminum paste in one gallon of unthinned dope.) This aluminum paste is made up of microscopic-size flakes of aluminum metal in solvents which carry and hold it. Mix the clear dope into the paste, not the paste into the dope.

Aluminum powder has been used extensively in the past, but because of the problems of handling the powder it has been superseded by the paste. If you should have powder, open the can slowly and carefully, so the powder will not float out into the air; then add some dope thinner and mix it into a paste before you remove it from the can. Mix this with clear dope in the same proportions as the paste; thin as necessary for spraying.

Sand the clear dope with number 400-grit paper and wipe the surface with a tack rag. Spray on one good heavy cross-coat of the aluminum dope. It should be noted that aluminum dope is not a sanding coat, but is used for ultraviolet protection. Too much aluminum powder will prevent a bond between the topcoats and the base coats. After the aluminum dope has thoroughly dried, wet-sand it just enough to get a smooth surface. Since the aluminum dope is used to protect the fabric from the sun's rays, it must not be sanded off. After sanding, it should completely cover the surface. This can be checked by shining a light inside the surface to find any thin spots where the light might shine through. Wash with clear water, using a sponge, and allow to dry thoroughly before applying next coat.

4. Finish Coats

After you have determined that the aluminum dope has formed a lightproof film over the surface, wet-sand it smooth, thoroughly bathe it down with water, sponge it off, and dry completely. Spray on a good cross-coat of the proper color butyrate dope. The pigments in butyrate dope are soluble in the solvents used in its manufacture, and the color goes right up to the surface. The gloss of the finish is then determined by smoothness of the film. To get a real good gloss, retarder is often mixed with the dope in the color coat to allow a longer time to flow out and therefore to form a smoother film. Another way, which is less costly, is to use thinner, wetter coats of the colored dope, thinned with regular thinner.

The finish described here is that used when you must balance between quality and cost. It will give adequate tautening and protection without the labor expense involved in the multi-coat, hand-rubbed finish so popular among home builders and antique restorers — whose sanding, spraying, rubbing, and polishing are all a labor of love. If you should be involved in an elaborate finish of this type, you should use as many thin, wet coats of butyrate dope as you feel necessary, wet-sanding with number 600 paper between each coat. Allow the finish to dry for at least a month (longer, if the climate is cold or wet), then compound the surface with rubbing compound and wax it. Compounding and waxing before the finish is completely cured can be detrimental to the dope. If a finish of this type is put on an airplane which was produced with the more conventional finish, you will have to determine whether the heavier finish has added enough weight behind the hinge line of the control surfaces that flutter could develop. Check the balance of the controls as recommended by the airplane manufacturer or by the local FAA district office.

Light overspray may be removed from the final coat by spraying it with a mixture of two parts of thinner and one part of retarder. This will melt the overspray into the finish and restore the gloss.

B. Polyester (Dacron®, Ceconite®, etc.)

A polyester fabric known as greige Dacron® or Ceconite™, a synthetic fiber which has not been passed over the shrinking rollers of the textile mill, is sold under several proprietary names for covering airplanes. This material, while having a longer life and greater tensile strength than Grade A cotton, must be installed on the authority of a Supplemental Type Certificate. It is available in more than one grade or weight, so be sure to use the grade specified in the STC instructions. The longer life of polyester may not always be advantageous, as the structure beneath will usually require attention before the fabric needs to be replaced.

1. Fabric Attachment

Inspect the structure to determine that all of the repairs have been properly made, all of the controls, wiring, and components which will be covered are ready, and all of the structure has been properly primed. Attach the polyester fabric to the structure with the heavy-bodied nitrate cement in the same way discussed under cotton and linen covering. It is put on smooth but not tight, and shrink with heat. A steam iron set on "wool" temperature may be

moved over the surface evenly, uniformly shrinking the material a small amount, several times. The total shrinkage should bring the fabric sufficiently taut with all of the wrinkles out. A heat gun or special heat lamp may be used instead of the iron, but be certain the temperature does not exceed 400°F. However, the iron method is preferred.

One beautiful feature of heat-shrinking polyester fabric is your ability to straighten sewn seams which have pulled into a curve in the application process. Apply a little local heat to the concave side of the curved seam until it straightens out; then uniformly heat the entire surface to get the proper tautness.

2. Prime Coats

While polyester fabric may be looked on as the answer to all problems as a covering material, it is not without its drawbacks. One of the main problems with this material is the difficulty of bonding the dope to the fibers. Organic fabrics such as cotton or linen are wet by the dope, and adhesion is not such a problem; but the inorganic fibers do not wet, and in order for dope to stick, it must completely encase or encapsulate the fibers. Butyrate dope does not do this as well as nitrate, so although the main dope film will be butyrate, for its fire-retardant qualities, the first coat should be a high-solids nitrate dope. There are several products on the market for use as prime coats for polyester fabric. These are essentially high-solids nitrate dope with some transparent pigments added for identity. These may not be called nitrate, but a simple test will verify their being nitrate, and thus having the required ability to encase the fibers. Rub a small amount of nitrate dope thinner on some of this dried primer material. If it softens the film immediately, it is nitrate and is good for the prime coat.

Thin this material in a ratio of about two to three parts dope to one part thinner, and carry only as much dope on the brush to the fabric as can be pushed through the fabric to surround and encapsulate every fiber. It should form a wet film on the inside, but be careful not to allow it to drip through to the opposite side of the structure. After this coat dries thoroughly, brush on a full bodied coat of nitrate, thinned only enough that it will brush on without pulling. Since polyester is not organic, there is no need for fungicides in the prime coats.

3. Fill Coats

After all of the rib-stitching has been done and the drain grommets, inspection rings, and surface tape have been applied, spray on the fill coats. Depending upon how tight the fabric is on the structure, you can spray on two full-bodied cross coats of either standard high-solids, or non tautening butyrate dope. "Non-tautening" is actually a misnomer, as all dope will tighten the fabric; but this dope has different types and amounts of plasticizers and will shrink the fabric less than standard dope. It is formulated to give the A&P more freedom in filling the polyester, and is normally tinted with a transparent pigment to identify it. Special dopes are sold for filling polyester fibers; these are usually high-solids butyrate, some having aluminum pigment in them.

A properly finished polyester fabric surface will still show the weave of the fabric. Since the polyester fibers are continually moving, any at tempt to completely hide them will result in a finish that does not have sufficient flexibility, except for 2.8 ounce polyester fabric.

After the clear butyrate dope has thoroughly dried, it should be sanded with about number 400-grit paper and the surface thoroughly cleaned with a tack rag, damp with a little thinner. Spray on one good cross-coat of aluminum- pigmented dope, made by mixing no more than 3-1/2 ounces by weight of aluminum paste to one gallon of un thinned butyrate dope. and thinned to proper spraying consistency. It should be remembered that the aluminum dope is not a sanding surface; it is used to protect the fabric and the clear dope from the ultraviolet rays of the sun. Polyester is not as susceptible to damage as cotton, but it is affected to a degree.) It should completely cover the surface, but not be so thick that the color coats lose their bond to the clear dope. Wet-sand with number 400-grit paper, wash the surface down with water, and sponge it off. Dry it thoroughly before spraying on the final coat.

4. Finish Coats

Spray on about three coats of butyrate dope, thinned for proper spraying. Here, as with cotton or linen, a good gloss color can be obtained with multiple thin, wet coats.

Conventional enamels or acrylic lacquers are not recommended over a doped surface because of the incompatibility of the materials. There is usually a rapid deterioration of the finish with such a system, and patching or rejuvenation is difficult or nearly impossible. Patching difficulties may be accepted by some users, however, in exchange for the chemical resistance and durable gloss provided by a polyurethane enamel.

If a polyurethane finish is desired over fabric, the finish is taken all the way through to one color coat in the conventional butyrate system, and then the polyurethane is applied as it would be over metal. If it is desired, a coat of butyrate dope the same color as the polyurethane may be sprayed on as a base for the topcoat. It has been found that the best adhesion may be had by the polyurethane if it is applied after 48 hours after the last coat of butyrate is sprayed on.

When a polyurethane-coated fabric needs patching, all of the finish must be sanded off, down to the aluminum dope and the conventional fabric patch put on as recommended by AC 43.13-1A, Chapter 3, Section 3. The finish is brought back up in the conventional butyrate system through the color butyrate and the polurethane finish is restored.

C. Polyester Over Plywood

An extremely smooth finish for plywood-covered aircraft structures may be acquired by covering with a lighter weight polyester fabric than is used for ordinary truss structure covering.

A couple of coats of nitrate dope are sprayed on the surface to provide a bed-coat for the fabric. After this dries completely, the fabric is pinned to the trailing edge, and a coat of thinned nitrate dope is brushed onto the surface of the wood. The fabric is laid onto this wet dope. A coat of nitrate is brushed over the fabric an the fabric smoothed to the surface. After this dries, two more coats of nitrate or butyrate are sprayed on, allowed to dry thoroughly, then sanded. A good coat of tautening butyrate dope is then sprayed on and sanded, then a cross-coat of aluminum-pigmented butyrate. The surface is wet-sanded, and after all of the sanding residue is washed off and the surface dried, the color coats of butyrate are sprayed on.

D. Fiberglass Cloth

Fiberglass cloth has an extremely loose weave, and is difficult to apply to an aircraft structure, but some of this has been pre-treated with butyrate dope and is used for aircraft covering with the authorization of a Supplemental Type Certificate.

1. Fabric Attachment

Before covering with fiberglass, the entire structure must be thoroughly inspected and all sharp edges removed or covered with tape. This is one of the "long life" fabrics, so the structure should have the best primer available applied — a good epoxy primer.

Because of the loose weave of the fabric and the butyrate pre-treatment of the fabric, attachment to the structure with full-bodied clear butyrate dope will provide adequate adhesion. The nitrate cements that are commonly used for cotton, linen, or polyester fabric cannot be used because of the butyrate pre- treatment. The nitrate will not penetrate the butyrate to give sufficient bond.

2. Prime Coats

The loose weave of the fabric and the butyrate pre-treatment make it almost impossible to brush the first coat of dope onto the surface. Clear butyrate dope should be sprayed on, thinned only enough that it will atomize properly. The air pressure on the gun should be low enough that the dope will not be blown through the fabric.

The first time you spray dope into fiberglass cloth, a test panel should be made and the technique perfected before trying it on the airplane. The dope should be heavy enough to thoroughly wet the fabric and soften the butyrate in the fabric, yet not so wet as to cause the dope on the back side of the surface to run. The glass filaments of which fiberglass cloth is made naturally cannot shrink. Tautening butyrate dope pulls the weave closer together to provide the smooth, tight surface. After this coat has dried, spray on a second coat, heavier than the first. Continue to spray coats of clear butyrate dope until the weave fills, and the fabric pulls up taut and wrinkle free. Brush on clear dope in which to lay the surface tape, drainage grommets, and inspection rings; then brush a coat of dope over all of them.

3. Fill Coats

After the fabric is taut and the weave full of dope, brush on two full-bodied coats of butyrate dope and allow them to dry thoroughly; then dry-sand the surface. Any time you sand the dope on the fabric-covered surface, you must be very careful not to sand through the fabric. Every stringer, rib-stitch knot, or screw head makes an easy spot for the abrasive to cut through.

Fiberglass is an inorganic fabric and will not be damaged by the ultraviolet rays of the sun. The clear dope can be damaged, however, so a coat of aluminum dope should be sprayed on and wet-sanded smooth. Any time aluminum dope has been sanded, be sure to wash the sanding residue completely off with water and scrub it with a sponge or rag; then dry the surface thoroughly.

4. Finish Coats

Several thin, wet coats of colored butyrate dope will allow the surface to flow out smooth, and a glossy finish will result.

Any time a fabric is applied according to a Supplemental Type Certificate, all materials used and all of the work done must be *exactly* as specified in the instructions which are part of the STC approval.

E. Built-Up Fiberglass Structure

Engine cowling, wheel fairings, and wing tips are often made of layers of fiberglass cloth embedded in an epoxy or polyester resin. These may be finished with acrylic lacquer or polyurethane enamel. Any time it becomes necessary to remove the finish from any such parts, it must be sanded off or taken off with acetone. Never use a prepared paint stripper. The active agents in the stripper will attack the resins and soften the component. When refinishing built-up fiberglass, fill any surface imperfections with spot putty and a sanding surfacer compatible with either enamel or lacquer sprayed on and sanded smooth. If the surface is to be finished with polyurethane enamel, polyester body filler such as is used in automobile body shops or boat repair shops will provide a good smooth surface for the epoxy primer and the final finish.

F. Radomes

Radomes are special fiberglass, honeycomb covers over the radar antenna. This material must be strong enough to withstand all of the airloads and ice accumulation likely to be encountered, and it must be electrically transparent. This means that it cannot shield or restrict the transmission or reception of the electrical energy from the radar equipment. No paint having metallic pigment can be used on a radome. Nonmetallic acrylic lacquer or polyurethane enamel are entirely satisfactory for this application.

G. Repair To Fabric Surfaces

Advisory Circular 43.13-1A describes in detail approved methods of repairing damaged fabric covering. These methods should be followed in detail, and after the repair is made, the finish should be brought up to the topcoat in the same way as the original finish.

It must be remembered that if polyurethane is used as the topcoat over fabric, every bit of it must be removed by sanding before the fabric can be patched. Touch-up may be done to polyurethane only if the gloss on the surface is broken by sanding it with number 400 or number 600 sandpaper. Once this surface has been roughened by the paper, the gloss cannot be restored by any method other than another coat of polyurethane. It will adhere to the sanded surface, and the application of a second coat should give no problem.

H. Repair To Fabric-Covered Plywood Surfaces

The wings of the Bellanca Viking are covered with mahogany plywood over which is placed a layer of polyester fabric. Before the application of the fabric, the entire wing is treated with a rot-inhibiting sealer and aircraft spar varnish. If a structure of this type becomes damaged, it should be repaired according to AC 43.13-1A, Chapter 1, Section 1. Before cutting the scarf on the patch material, coat both sides with rot-inhibiting sealer and spar varnish and allow it to cure. Cut the scarf and complete the repair. Sand the patch smooth, brush on one coat of nitrate dope and allow it to dry. Brush on a second coat and lay the fabric patch into this wet dope. Immediately brush on another coat of nitrate dope, and while it is wet, press the fabric down to the surface. Allow this dope to dry completely and sand it. Apply the normal build-up, aluminum-pigmented coats, and final finish to match the remainder of the airplane.

QUESTIONS

1. Where can a person find detailed, authoritative instructions for covering an airplane with polyester fabric?

2. Who should aircraft fabric be given a pull test before being used to cover an airplane?

3. Who should an airplane freshly covered with Grade A cotton be doped as soon as possible?

4. Who is nitrate better for a prime coat than butyrate?

5. When is the fungicidal treatment applied to cotton fabric?

6. What is the purpose of the aluminum coat in a dope finish?

7. How tight should polyester fabric be shrunk by heat?

8. Why should nitrate dope be used as a first coat for polyester fabric?

9. How long should you wait after the last coat of Butyrate is sprayed on polyester fabric before you spray on polyurethane?

10. Who should you spray the first coats of dope on fiberglass cloth?

11. How can dope shrink fiberglass cloth?

12. How may old finish be removed from a fiberglass wheel fairing?

13. What kind of paint may not be used on aradome?

CHAPTER III

Application Procedures

A. Spray Painting

1. Gun Operation

Both the pressure-fed and suction cup guns have three valves to give the operator control of the film he is applying. The air valve, Figure 4, is opened when the trigger is first pulled. This allows air to flow out through both the atomizing holes and the wing port nozzles. The only control of atomizing air is by adjusting the pressure at the regulator, but the spray width adjustment controls the amount of air allowed to flow through the wing ports.

Continued pulling of the trigger opens the fluid valve so the material can be either forced out by the pressure on the pot or pulled out by the low pressure developed at the fluid nozzle by the atomizing airflow. Adjusting the material control valve determines how

much this valve will be pulled off its seat and how much fluid will flow.

As the fluid leaves the nozzle, it is broken up into tiny droplets by the atomizing air and is sprayed on the surface being finished.

If no air is allowed to flow through the wing ports, the spray pattern will be circular. As the spray width adjustment screw is turned to the left, the valve opens further, allowing the air from the wing ports to flatten the spray pattern. As the width of the spray is increased, the material valve must be opened further to get proper coverage for the larger area.

When paint is sprayed from a pressure-fed gun, the pressure on the pot determines the amount of fluid which will be sprayed. The biggest mistake in using this type of equipment is getting too much material and having runs and sags in the finish. The

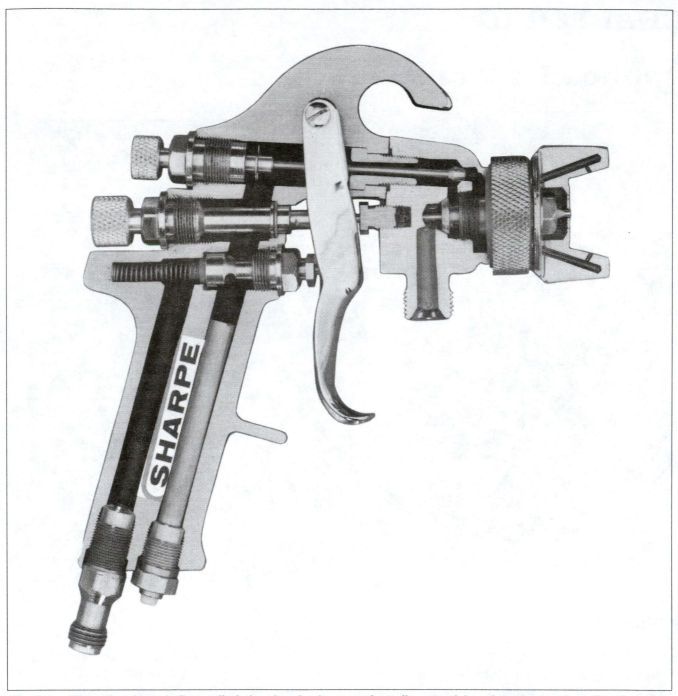

Figure 4. *When the trigger is first pulled, the air valve is opened sending atomizing air to the nozzle and to the wing ports. Continued pulling of the trigger pulls the fluid needle off its seat and allows material to flow from the nozzle. The fluid adjustment screw determines the amount of fluid allowed to flow, and the aira valve controls the shape of the spray pattern.*

air pressure on the pot should be low enough to get just enough material to do the job, and then you need only sufficient air pressure on the gun for proper atomization. A good way to determine the correct pressure is to begin with 35 to 40 psi on the gun and bring the fluid pressure up to match the air, rather than bringing the air pressure up to match the fluid.

There should never be more than about 10 psi on the pot unless there is excessive line loss in the hose. Six or eight psi on the pot is enough for most acrylic lacquers. There should be just enough air to atomize the fluid properly. The use of low pressures prevents air impingement, sags, and runs. This generally produces good wet coats which flow out smooth.

18

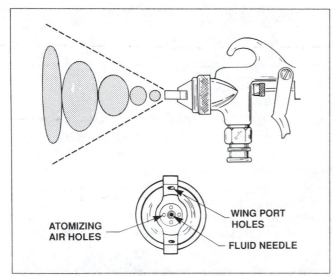

Figure 5. *The amount of air allowed to flow through the wing port holes determines the shape of the spray pattern. As the spray pattern width is increased, the material valve must be opened more to get proper coverage for the larger area.*

Pressure on a pressure cup or suction cup gun can vary from 20 to about 55 psi, depending on the equipment and the operator.

When spraying acrylic lacquer, be sure the material is thinned sufficiently. There should never be more than four parts of color to five parts of thinner, and when a suction cup gun is used, the proportion of four parts of color to six or seven parts of thinner is more reasonable.

2. Spray Gun Malfunctions

a. Spitting

Spitting, or interruption of the fluid flow, is caused by air getting into the fluid passageway.

Figure 6. *The air pressure on the pressure pot determines the amount of fluid that will be delivered to the gun.*

On a suction cup gun, this could be caused by a dried out packing around the material needle valve, C of Figure 7, allowing air to get into the fluid passageways. This can be remedied by lubricating the packing with a few drops of light oil. Dirt between the body of the gun and the fluid nozzle seat, D of Figure 7, will allow air to enter at this point. Finally, a loose or defective nut attaching the gun to the suction cup, E, could allow air to enter the fluid stream.

b. Distorted Spray Pattern

The normal spray pattern is fan shaped, A of Figure 8. The long axis of the fan is perpendicular to the wing ports. The width of the fan is determined by the amount of air allowed to flow out of the wing ports. When the width of the spray pattern is increased, the amount of material must be increased to get proper coverage. If a situation exists where you cannot handle a large spray pattern, while spraying inside of a wheel well or other restricted area, for instance, you must cut down on the amount of fluid pressure. In order not to distort the spray pattern, the amount of atomizing air must be decreased accordingly.

A spray pattern which is basically fan shaped but is heavy in the middle, B of Figure 8, could indicate

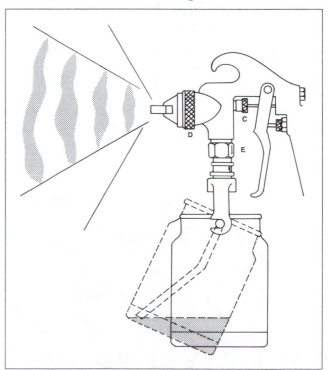

Figure 7. *Paint spitting may be caused by air entering the material line. This can be because of too low a fluid level in the cup, by a leaking packing (C), or attaching nut (E), or by a poor fit of the fluid nozzle (D).*

insufficient atomizing air pressure. Increasing the air pressure to the gun will correct this situation.

The opposite condition — a dumbell-shaped pattern, C — occurs when you have too much atomizing air pressure, or you are attempting to get too wide a pattern with thin material. This may be corrected by increasing the amount of material and decreasing the amount of air from the wing ports.

A pear-shaped pattern, D, indicates that there is probably a buildup of material around one side of the fluid nozzle, cutting off the flow of atomizing air to one side of the pattern. Remove the air nozzle and soak it in thinner to clean it out. Do not scrape or probe it with wire or a metal scraper, as you will scratch and damage these passages. A damaged or loose-fitting air nozzle will also cause this type of distortion.

A crescent- or banana-shaped pattern, E, indicates one of the wing portholes is plugged up, allowing the pattern to be blown to one side. Remove the air cap and soak it in thinner and blow the passages out with compressed air.

3. Spray Gun Cleaning

As with any precision tool, a spray gun will give a long life of satisfaction if it is properly maintained, but will give dissatisfaction if it is not cared for.

The gun must be kept clean. If a suction cup is used, immediately after spraying, dump the material from the cup and clean it. Put some thinner in it and spray it through the gun. Trigger the gun repeatedly while spraying the thinner; this will flush the passageways and clean the tip of the needle. Spray until the thinner comes out with no trace of the material.

When cleaning pressure-fed guns, first empty the gun and hose back into the pot. Loosen the air cap on the gun and the lid of the pressure pot. Hold a

rag over the air cap and pull the trigger. Atomizing air backing up through the gun and the fluid line will force all of the material back into the pot. Empty and clean the pressure pot; then put thinner in it and replace the lid. Spraying thinner through the hose and gun will clean the entire system.

After the inside passages of the gun are cleaned, soaking the nozzle in a container of thinner will further clean the head. Do not soak the entire gun in the thinner as this will ruin the packings. The air valve stem and all of the packings around the fluid needle should be lubricated with light oil so they will operate smoothly and the packings remain soft and pliable. The packing nuts should be tightened finger-tight only.

Material should never be left in the gun, as it will set-up and plug the passages. If the passages become plugged with dope or acrylic lacquer, they may be cleaned by disassembling the gun and soaking the parts in acetone or MEK. Catalyzed materials such as epoxies and polyurethanes, if not flushed out immediately after use, will set-up in the gun and hoses. When this happens, the hoses must be discarded and the passages in the gun cleaned by digging the material out. This is not only time consuming, but there is a good probability that the inside of the gun will be damaged.

4. Spray Technique

The most important considerations in spray painting are the use of the proper gun, fluid tip, and needle, and the proper air pressures and fluid viscosity for the material being applied. Once these have been selected and adjusted, the final determining factor in the quality of a paint job is the application procedure.

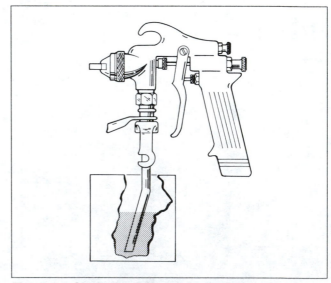

Figure 9. *Clean the spray gun immediately after using by spraying clean solvent through it until there is no indication of the material.*

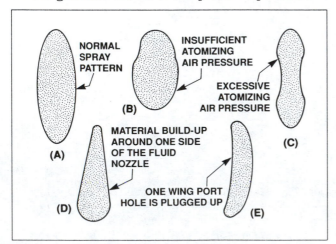

NORMAL SPRAY PATTERN

INSUFFICIENT ATOMIZING AIR PRESSURE

EXCESSIVE ATOMIZING AIR PRESSURE

MATERIAL BUILD-UP AROUND ONE SIDE OF THE FLUID NOZZLE

ONE WING PORT HOLE IS PLUGGED UP

(A) (B) (C) (D) (E)

Figure 8. *Paint spray pattern defects.*

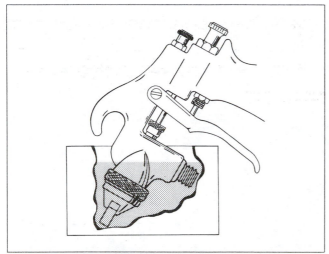

Figure 10. The spray gun head may be cleaned by soaking the nozzle in a container of thinner. CAUTION: Do not soak the packing around the fluid needle.

The nozzle of the gun should be held between six and ten inches from the surface, depending on the material, Figure 11. It should be close enough to lay a good wet coat on the surface, yet far enough away that the material does not run or sag.

The gun should be held perpendicular to the surface so the material will spray out in an even pattern. If the gun is tilted or tipped, Figure 12, the pattern will be heavier on the side nearest the gun, and dry and rough on the side farthest from the gun.

Move the gun parallel to the surface being sprayed. Begin the stroke, then pull the trigger. Release the trigger before completing the stroke, Figure 13. If the gun is arced when spraying, the surface will be uneven; there will be a heavy deposit where the gun was nearest the surface and a thin one where the spray arced away.

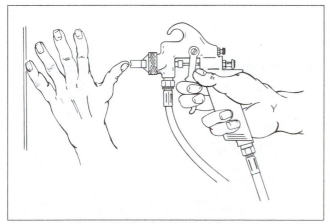

Figure 11. The nozzle of the spray gun should be held between six and ten inches from the surface being sprayed.

Before starting to lay the film of paint over the flat part of the structure, cut in the edges and corners. This is done by spraying along the corner which gives the thickest coat along the edge and blends out in the flat portion, Figure 14.

A single layer of material laid on the surface by one pass of the gun will be typically about 10 to 12 inches wide, thicker in the middle and taper off at each end. In order to get a good, even build-up of finish, spray on the first pass; then come back with the gun on the return pass, overlapping all but about two or three inches of this first pass. The third pass will overlap all but about two or three inches of the second. Continue this overlap and the resulting finish will be a nice even film with no runs or sags.

B. Aircraft Painting Sequence

Anything we do in aviation maintenance must be carefully planned if we expect to produce effectively. In painting an airplane, considerable planning should precede the actual shooting. The airplane should be positioned in the booth in such a way that the airflow will be from the tail toward the nose so that you can paint in this direction and the overspray will be ahead of you.

If it is possible, have two painters work simultaneously on opposite sides of the airplane, working away from each other. In this way, the overspray problems will be minimized.

First, paint the ends and leading edges of the ailerons and flaps; then, the flap and aileron wells, the wing tips, and leading and trailing edges. Spray all of the landing gear, the wheel wells, and all of the control horns and hinges. In short, before starting on any flat surfaces, paint all of the difficult areas, then proceed in a systematic sequence.

Paint the bottom of the airplane first, using a creeper for the belly and the bottom of low-wing airplanes. Prime the bottom of the horizontal tail surfaces first, starting at the root and working outward, spraying chordwise. Then work up the fuselage, allowing the spray to go up the sides. Work all the way up to the engine. Spray the bottom of the wing with each painter starting at the root and working toward the tip, spraying chordwise.

Jack up the nose of the airplane to lower the tail enough to allow the top of the fin to be reached. Both painters work together with one slightly ahead of the other so they will not spray each other. When spraying the top of the fuselage, tilt the gun so the overspray will be ahead of the area being painted and the new material will wipe out the overspray. The primer should be sprayed across the fuselage, and

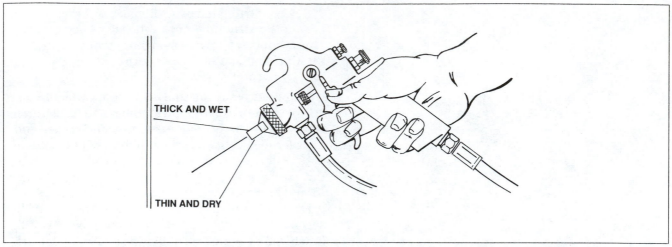

Figure12. If the gun is not held perpendicular to the surface, one side of the spray pattern will be wet with a tendency to run, and the other side will be rough and dry.

spanwise on the vertical and horizontal tail surfaces and the wing.

After the primer has cured for the proper time and is ready to receive the top coats, the same sequence is used to spray on the finish. The tack coat is sprayed on the bottom surfaces starting at the center of the fuselage and spraying across it, then out the horizontal surfaces spanwise. The top of the aircraft has the tack coat sprayed lengthwise on the fuselage and chordwise on the surfaces.

The final coat is sprayed on, using the same sequence and direction as the prime coat. The bottom of the fuselage is sprayed crosswise and the wing and tail surfaces are sprayed chordwise. The top of the airplane is sprayed across the fuselage and spanwise on the wing and tail surfaces.

It is often impossible to reach completely across the top of the wing, so spray as far as you can reach while working from the root to the tip, along the trailing edge; then walk around the tip and work

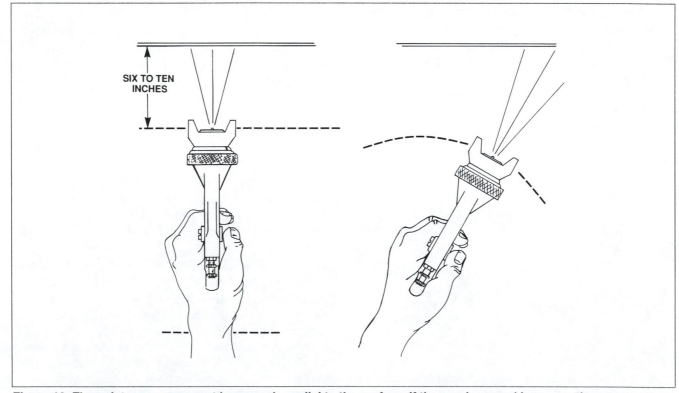

Figure 13. The paint spray gun must be moved parallel to the surface. If the gun is moved in an arc, the spray pattern will be thick in some spots and thin in others.

back toward the fuselage. Keep the gun tilted back so the overspray will not fall on the rear half of the wing where the paint has hardened to such a point that the overspray will not blend in.

Spraying on a coat of acrylic lacquer with an excess of solvents can be used to wash out acrylic overspray. This softens the film and allows the overspray to sink into the finish. Dried overspray from any material other than polyurethane can be "burned down" or "washed out" by spraying a mixture of one part retarder and two parts thinner on the surface while the overspray and base finish are still fresh. This mixture will soften the surface enough to allow the overspray to sink in and allow the surface to gloss. Enamel overspray does not usually present the problems of lacquer or dope, since it has so much slower drying rate. The overspray can sink into the finish while it is still wet.

C. Painting Safety

The need for safety procedures can really shock us when we see a hangar, complete with airplanes, go up in smoke, all because we didn't observe a few basic facts about safety. Hangars can be fire traps and airplane maintenance hazardous, especially when we are doing painting or fabric work.

Take the business of sanding a wing being recovered, for example. Here, you have a set of conditions which could give you some real excitement.

Let's see what is happening: The dope sometimes used for the first coat over cotton fabric, and almost always for the first coat with polyester is nitrate,

preferred over butyrate because it encapsulates the fibers better. Some of the special primers used for polyester may not be identified as nitrate, but that is what they usually are. (You can check this by rubbing a little bit of nitrate dope thinner over a dried area covered with the primer. If the finish is immediately softened, it is nitrate.) Nitrate dope is made from nitrocellulose and certain solvents, and nitrocellulose in the manufacturing process is naturally explosive.

When the wing is being doped, supported on padded saw horses, the fumes from solvents and the nitrocellulose fill the inside of it. After the dope dries, you grab a piece of dry sandpaper and go after the roughness on the surface. Now, when you wipe or rub across a non-conducting surface with another non-conductor, the same thing happens as when you slide across plastic seat covers: you generate a static charge on the surface.

If you are wearing rubber-soled shoes, your body has the same electrical charge as the wing, and nothing happens—yet. But, if you are called away for a few minutes and get a drink at the drinking fountain, or touch the hangar structure, the electrical charge on your body flows off to the ground, and you are electrically neutral. Then, when you get back to the wing, if you should, let's say, move the control cables sticking out of the root, the static charge on the wing jumps to the cable, sparking inside the wing, and flows to you. This spark is all that is needed to ignite the explosive fumes inside, and you have your hands full of trouble—real quick!

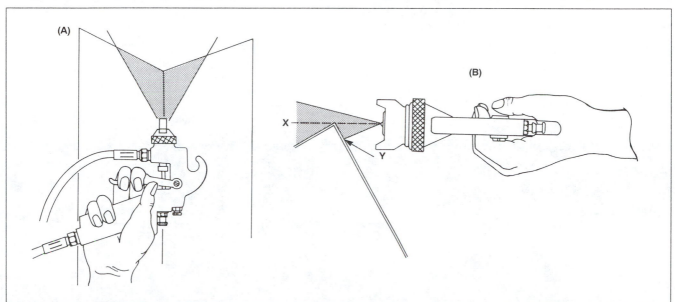

Figure 14. A   When cutting in cornerrs of a surface, spray parallel to the corner first, then blend this stroke into the rest of the surface by spraying perpendicular to the corner.

Figure 14. B   Fan will not go around corner, but will have heavy build-up and runs at "Y" and bare spots at "X".

The best way to prevent all this is to eliminate the static voltage difference: **GROUND THE SURFACE YOU ARE SANDING.** Run a fairly heavy wire from the surface to a good ground source, such as a cold water pipe, or to a bare part of the metal hangar structure. In this way, the surface will be at ground potential while it is being sanded, and you will not build up a voltage on it.

Any time you are working around materials which release explosive vapors, or any time static electrical charges may present a problem, wear cotton clothes. Most of the synthetic fibers are more inclined to build up a static charge than cotton. Wear leather-soled shoes so you will be grounded and will dissipate any charge that builds up, rather than carrying it. When you spray dope or lacquer, it is important that the airplane, the pressure pot, the hose, and you, yourself, all be grounded together.

Since butyrates, acrylics, and polyurethanes do not use nitrocellulose, they do not create so much fire hazard; but dried overspray from nitrate dope can be a real danger on the paint booth floor, and any time you must sweep it up, WATCH OUT!! Dried nitrate dope overspray is highly explosive, and sweeping will create enough static electricity to ignite it. To clean the floor, douse with water and wet-sweep.

If the floor of the spray booth has dried nitrate or butyrate dope overspray on it, be sure it is removed by wet-sweeping before allowing any zinc chromate overspray to mix with it. If these oversprays are mixed, spontaneous combustion can result, and you can get a dandy fire.

The overspray from certain enamels, if swept up and put in a pail of water, can catch fire by themselves. Rags and sponges which have been used to apply one of the phosphoric acid conversion coatings such as Alodine should be washed out thoroughly before being thrown away. If the material is allowed to dry in the rag, there will be a danger of it catching fire from spontaneous combustion.

All overspray residue should be kept in covered containers away from the buildings where spraying is done.

Mixing dopes or lacquers also can be a hazardous process; it is best done on a paint shaker. Mechanical stirring with a rod and blade on an electric drill motor is **DEFINITELY NOT RECOMMENDED.** You will do a lot of pounding and think you are doing a good job mixing the paint, but you really are not. What you are really doing is stirring up a lot of fumes which rise around the arcing drill motor, and, if the atmospheric

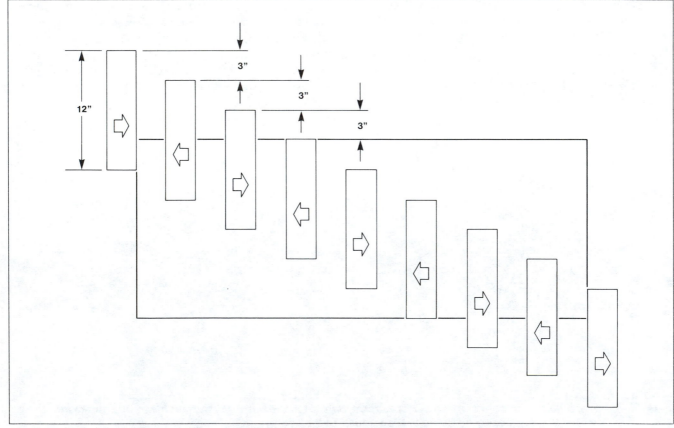

Figure 15. When building a paint film, overlap each pass by all but about two or three inches.

conditions are right, will start a flash fire in the top of the paint can. What invariably happens then, is that someone will try to carry out the can of burning paint, and spill it; this creates lots of excitement!

If you should ever get a fire in a can of paint, immediately cover the can; drop the lid back on it, use a piece of cardboard, or even a cloth whatever is handy. Almost any kind of cover will either smother the fire, or at least contain it, until you can reach the fire extinguisher.

Another safety factor that goes without saying is the importance of proper air movement in the spray area. A properly designed spray booth has an air movement system that not only keeps the air circulating, but removes all of the solids and solvents. Since all the materials used in painting are heavier than air, the exhaust system for a booth should be near the floor. If you spray in an area not designed primarily as a spray booth, you should at least be sure there is enough air movement to leave no more than a mild odor of the finish material while you are spraying. A heavy concentration of fumes is dangerous, not only as a fire hazard; excessive concentration of fumes will deplete the oxygen supply required by the operator.

Modern aircraft paints and dopes do an excellent job of extending the life of an aircraft structure, as well as making it a lot more pleasing to look at. They do require proper handling and application the same careful attention to detail we accord any phase of aircraft maintenance. Good operating procedures are safe operating procedures.

1. Polyurethane Paint Safety

The use of polyurethane paints requires that certain safety precautions, attention to health hazards, and medical surveillance be observed and that protective equipment be worn. The following information is taken from United States Coast Guard Aviation Technical Note 2-68B.

a. Safety Precautions

During transit and storage of polyurethane paints, a safety hazard exists when a defective batch of resin undergoes slow deterioration as a result of moisture contamination. This resin defect manifests itself in the form of a bulging can, by the emission of other than the normal odor, or by a change in the resin from a clear to a cloudy state. This defect will result in a slow build-up of carbon dioxide, the pressure of

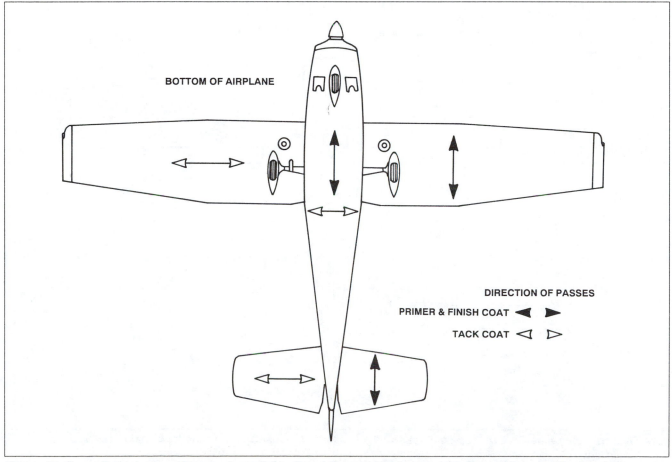

Figure 16. Spray the primer and topcoat parallel to the airflow and tack coat spanwise and across the fuselage.

25

which could cause the paint can to burst with sudden explosive violence. To minimize complications which may result from resin defects, paint shop personnel should periodically inspect incoming defective resins as outlined above. All defective containers of polyurethane paints for signs of material as described above should be removed and disposed of with caution, and the incident appropriately reported.

When painting a complete aircraft or working in confined areas, adequate ventilation and/or appropriate facemask breathing protection should be provided to minimize toxicity effects.

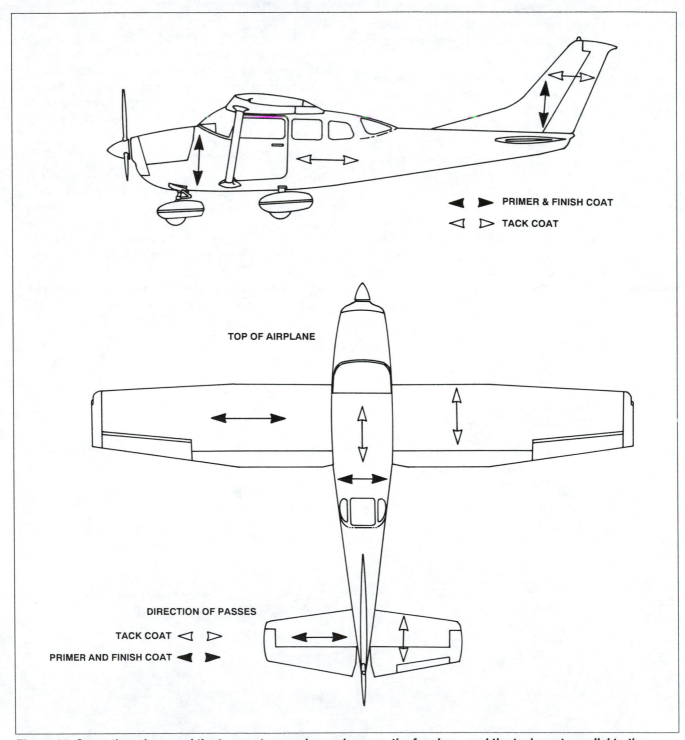

Figure 17. Spray the primer and the top coat spanwise and across the fuselage, and the tack coat parallel to the airflow.

b. Health Hazards

Polyurethane paints can produce irritation of the skin, eyes, and respiratory tract during mixing and application. Allergic sensitization of personnel exposed to the vapors and mists produced during spray application may occur and cause difficulty in breathing, dry cough, and shortness of breath. Individual susceptibility appears to be a controlling factor. Once sensitized, many workers cannot tolerate even a minimum subsequent exposure and must thereafter avoid work areas where such exposure could occur.

c. Medical Surveillance

For production type painting, medical surveillance as described below should apply to personnel mixing or applying the paint.

Selected persons shall receive a medical evaluation prior to assuming these tasks. Persons found to be medically qualified and assigned to perform these tasks will be reevaluated at specified intervals. The medical evaluation shall include but not necessarily be limited to the following:

(1) A complete medical history to exclude personnel with any cardiac ailment or respiratory disease, e.g., asthma, emphysema, or respiratory allergies.

(2) A physical examination shall be repeated annually.

(3) A complete blood count and chest X-ray (14" x 17") shall be obtained initially and repeated annually.

(4) Pulmonary function studies to include a one-second forced expiratory volume (FEV-1) shall be conducted initially and repeated semi-annually.

Personnel involved in painting operations who display any of the symptoms described in Health Hazards shall be removed from the painting assignment.

d. Protection Equipment

Production type mixing and spray painting operations shall be conducted in specially designed, exhaust-ventilated areas, using personal protective equipment as follows:

(1) A well-fitted, double cartridge organic vapor respirator with fresh cartridges inserted daily.

(2) Solvent-resistant gauntlet style gloves.

(3) Safety goggles.

Painters should be fully clothed with collars buttoned and sleeves taped at the wrist.

QUESTIONS

1. What does the painter adjust to vary the amount of atomizing airflow?

2. What determines the shape of the spray pattern?

3. What is the maximum amount of pressure you should use on a pressure pot for normal painting?

4. How much should acrylic lacquer be thinned for spraying when used in a suction-cup type gun?

5. What would be likely to cause a fat, fan-shaped spray pattern?

6. What would be likely to cause a dumbbell shaped spray pattern?

7. What would be likely to cause a pear-shaped spray pattern?

8. What would be likely to cause a banana- or crescent-shaped spray pattern?

9. Why should a spray gun be triggered while cleaning by spraying thinner through it?

10. Why is it not recommended to soak the entire spray gun in thinner to clean it?

11. How can hardened dope be removed from the internal passages of a spray gun?

12. How can hardened polyurethane be removed from the internal passages of a spray gun?

13. How far from the surface should the nozzle of a spray gun be held?

14. What type of finish will result if the spray gun is tilted while spraying?

15. How much should each pass by a spray gun overlap the previous pass?

16. Why is it advantageous to have two painters working on an airplane simultaneously when spraying it?

17. Which should be painted first, the flat area, or the more difficult irregular areas?

18. What can be done to "burn down" acrylic over-spray on an airplane surface?

19. Why is enamel overspray on a surface less of a problem than lacquer overspray?

20. Why should a doped surface be grounded when dry-sanding it?

21. Why is it recommended that leather-soled shoes be worn in a paint shop?

22. How can dried nitrate overspray be safely swept up?

23. Why is it definitely not recommended to use a stirring paddle on an electric drill motor to stir paint?

24. Should the exhaust fan in a spray room be located near the floor or higher up the wall?

CHAPTER IV

Finishing Problems

A. Filiform Corrosion

Any airplane finish, especially polyurethane, which has indications of puffiness under the film, primarily around the faying strips, can be a warning of real trouble. If the polyurethane was applied over wash primer which had not cured because of insufficient water in the air, it will trap the still active acid against the skin. The almost impervious film of the polyurethane will allow only a small amount of water to penetrate; not enough to finish converting the acid, but enough to form corrosion salts with the aluminum skin.

If you pick the film in these puffy areas and find them full of dry powder, you most probably have a good case of filiform corrosion. This is actually a common type of corrosion which gets its fancy name from its filament-like formation. If any trace of corrosion is found, it is quite likely to be widespread, and by its nature, corrosion will continue to eat the metal of an airframe as long as any corrosion products are allowed to remain on the surface. As drastic as it sounds, if corrosion is found, the only reasonable action to take is to strip all of the finish from the airplane and remove every trace of the corrosion.

Methods of corrosion removal and the treatment of the metal surfaces are covered in the training manual *Aircraft Corrosion Control* and in the FAA advisory circulars 43.4 and 43.13-1A.

When all of the paint is off the surface, remove the corrosion products by scrubbing them with aluminum wool or a nylon scrubber, such as 3M's Scotch-Brite. Do not use steel wool, as the steel particles will embed in the aluminum and cause more corrosion than was originally on the surface.

When all of the corrosion products have been removed, the surface should be carefully examined to assess the damage. If no skin needs to be replaced, the entire surface should be treated with a conversion coating such as Alodine, to form a protective phosphate film on the surface. Apply this exactly as recommended by the manufacturer and then spray on an epoxy primer; then, finally, a new polyurethane finish.

Since the filiform corrosion forms because of uncured wash primer, the use of an epoxy primer under the polyurethane will prevent this expensive problem.

B. Color Matching

No matter how good a repair has been made, if the finish does not match the original, the customer is likely to be less than satisfied with the job. And, matching paint can sometimes be a real chore. Paint gets its color by reflecting light of the various wavelengths, and since different kinds of light are made up of different mixes of wavelengths, a color may look like a good match in the spray booth but really disappoint you when the airplane is rolled out into the bright sunlight.

When paint pigments are prepared in the laboratory to match a specific color, a sample may be run in a spectrophotometer and the results fed into a computer. This prints out, among other things, the metamerism index. This tells how well the colors will match under *all* light conditions. A well-designed pigment from a reputable paint manufacturer can be depended upon to give a good color match *if properly applied*.

Why, then, do we sometimes get a poor match? This can be because of either improper mixing or improper application.

Pigments in a finish are suspended in a liquid known as the vehicle, and if the can sits on the shelf for an extended period of time, the pigments will settle out, and they must then be thoroughly mixed before you can expect to get a really satisfactory color match.

Ideally, any pigmented material should be mixed on a mechanical shaker by putting the can in the shaker upside down, and shaking it for fifteen to twenty minutes. If you do not have access to a mechanical shaker, you can do a good job of mixing by following these steps:

1. Pour off half of the can of material into a *clean* can of the same size as the one you have just opened.

2. Stir or shake the remaining material until *every bit* of the pigment is in suspension. This is important with any finish, but especially with the metallics. Some of the metallic pigments may have as little as 1/10 of an ounce of some

components in five gallons of finish, and this little bit could collect in the seal ring around the bottom of the can. From this, you can see that stirring with a paddle or beating with a blade on a drill motor will leave a lot to be desired in adequate paint mixing.

3. Pour all of the paint from the first can into the second, and carefully examine it to be sure you have loosened every trace of the pigment from the bottom.

4. After you are certain that every bit of the pigment is in suspension, "box" the material by pouring it back and forth between the two containers until it is *thoroughly* mixed.

The undercoat of the finish has a lot to do with the final color match. Many of the pigments are transparent enough to pass some light which reflects from the base coat. Dope, enamels, acrylics and polyurethanes are color-matched over a white base coat. The high-solids content of polyurethane, in excess of 60% as compared with somewhere around one-half of that for acrylics, does not make them less critical as to their base coat color.

Even if a metallic finish is sprayed over a white base coat, we cannot be sure of a perfect color match every time. A drastic difference can result even from the same batch of paint, shot from the same gun, by the same operator, on the same background, at the same time. How can you, then, get a match? By varying one or more of these:

1. The spray pressure
2. The amount of thinner
3. The number of coats

If metallic material is applied wet and/or heavy, it will be dark and will have a tendency to be dull. If it shot on light or dry, it will be too light colored and too bright — too metallic looking. Changing the spray techniques or the air pressure will change the color. In order to match metallic colors, especially gold metallics, you just about have to do it by trial and error, but you have three variables to play with: air pressure, amount of thinner and number of coats.

About the best way to be assured of a good color match is to use the same paint as originally used; but if the surface you are matching has faded or you are not able to get exactly the shade you want, you can use the color wheel of Figure 18 and come up with a pretty good match.

Red, blue, and yellow are the basic colors, and when they are combined, they produce the green, orange, and purple color families. If, for example, you are matching an orange that is just a little bit deeper than the one in your gun, add just a few drops of red.

If it is too deep, it can be lightened with a few drops of yellow. The same applies to the green and purple families. If you go directly across the color wheel, you will get a gray. If, for example, you mix blue and orange, you will get gray. The relative amounts of blue and orange will determine, of course, whether the gray has a bluish or an orange cast.

After a color has been mixed to the hue or tone you want, it can be lightened by adding a little white or darkened by adding a few drops of real dark blue. If black is used to darken a color, it will have a tendency to become muddy looking.

Research in pigments for aircraft finishing has brought out some good, color-fast pigments which maintain their colors for years. In the past, some companies have used lead in some of the pigments for its non-fading characteristics, but modern organic pigments provide any color you want without the tendency toward chalking or fading.

There are three price categories of pigments used for aircraft finishes:

1. Blacks, whites, grays, and solid colors except reds
2. All metallics and reds3. Exotic colors

The color fastness or durability is about the same for all three categories, but certain ingredients or the expense of manufacturing pigments to provide certain special colors or shades cause their price to be higher than that of others; these are the exotic colors.

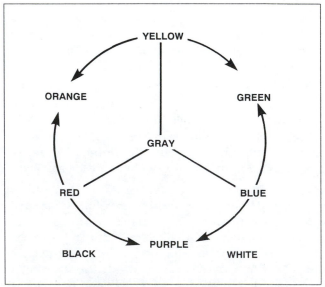

Figure 18. Red, blue, and yellow are the basic colors. Mixing yellow and blue produces green, blue and red make purple, and red and yellow give orange. All three mixed together will result in a gray.

Red-colored finishes are more costly than most of the other solid colors because of the special ingredients in their pigments. One type of red pigments, known as the "bleeding" reds, has given A&P's trouble for years when putting another color over it. The pigments used in bleeding reds are soluble in the solvents used in the topcoats and allow the red to bleed right on up through whatever is placed over it. The pigments in nonbleeding reds are soluble only in the manufacturing process and are not dissolved by the solvents used in spraying.

Several years ago, there was a lot of interest in vivid colors, designed to reflect a maximum amount of light and make the airplane more visible in smoke or haze. We still see this on some of the airplanes operated by the FAA. One of the main problems with this finish was its rapid fading. Today with the increased interest in making our airplanes as visible as possible, there has been a recurrence in the use of vivid colors. A vivid color coating is actually a transparent color coating; usually orange, red, yellow, or green applied over a good white reflective base coat. After the vivid coating has been applied and dried, a clear, ultraviolet-absorbing topcoat is put on. This helps prevent the sun's rays dulling the finish and helps these modern vivid colors last longer than the older ones.

C. Difficulties With Acrylics

The basic problem with the application of acrylic lacquer lies in its low solids content. The hiding quality of this material is poor, and the tendency is to spray it on too thick. If the lacquer is too viscous for proper spraying, excessive air pressure must be used.

An acrylic film sprayed from too thick a material may produce a glossy surface, but if you look across it, it will appear hazy. If you look at the surface with a magnifying glass, you will find millions of tiny holes. These are air impingement bubbles caused by air being introduced into the material by excessive atomizing air pressure. These tiny bubbles do not coalesce, or join up with others to form larger bubbles and make pinholes in the film; they remain tiny, and form the air impingement haze, or "stardust," as it is sometimes called.

To prevent air impingement, thin the acrylic lacquer at least in a ratio of four parts of color to five parts thinner. This may seem too thin, because of the low covering power of the material, but it is necessary to keep it thin in order to keep the atomizing air pressure low enough that no air impingement haze will be produced. Multiple, thin coats are to be chosen rather than fewer coats of thick material.

D. Difficulties With Polyurethanes

Its high-gloss, chemically resistant, tough, but flexible film seems to make polyurethane without peer. It is not entirely, however, without its problems. The superior hiding quality of polyurethane is due to its high solids content, and this very characteristic can give problems with runs and sags. A very light tack coat is sprayed on the surface and after the solvents flash off, about fifteen minutes, a full wet coat is sprayed on. This may not cover to your satisfaction, and cause you to spray on another coat. Since this material is so long flowing, this second coat is quite likely to cause it to sag or run. Spraying onto a cold skin when the air is warm will also likely cause a sag.

All pigmented materials have a tendency to settle out, and in addition to the thorough mixing before spraying begins, the pressure pot should have an agitator to keep the material moving all the time spraying is in progress. The agitation must be slow and constant. Fast agitation will mix tiny air bubbles into the material and they will be carried to the surface where they will produce the microscopic air impingement surface defects. This is like millions of tiny pinholes caused by air from the pot becoming entrapped in the surface.

Temperature affects the cure of polyurethane in a marked way; high temperatures cause a rapid cure, while lower temperatures allow a longer flowing-out time. It is desirable that the temperature of the metal be not much lower than 50° to 60° degrees Fahrenheit when spraying.

The humidity is also an important factor to be considered. High humidities are desirable as this accelerates the cure, but if the humidity is excessive, the finish will have a defect known as fuming. Here millions of microscopic bubbles form in the surface of the material and become entrapped in the finish.

An excessively heavy coat of finish will cause gassing in the curing process, and the surface will contain all of the tiny holes that result from this gas.

E. Difficulties With Dope

1. Fire Hazard

Everyone knows that nitrate dope is a fire hazard, and that butyrate is less hazardous because it will not support combustion. Polyester fabric with butyrate dope has enjoyed popularity, both because of its long life, and supposed fire resistant finish. Polyester is an inorganic material and does not absorb the dope; rather, the dope must wrap around or encapsulate each of the fibers. Butyrate dope does not do this as well as nitrate, and for this reason, it has been replaced as a prime coat by either straight nitrate dope or some of the proprietary first coat materials. These are usually nitrate dope with a little coloring for identification, and some special solvents. This, while providing a better bond than butyrate, does have the problem of the faster burning rate. When butyrate is put over nitrate, it affords but little fire protection. It retards the burning of the nitrate, but not significantly.

2. Dope Adhesion

The prime coat of nitrate provides a good bond to the fabric and the butyrate buildup and topcoats will bond to the nitrate. The solvents used in butyrate will soften the film base of either dope, but the nitrate solvents will not soften the butyrate base and therefore, nitrate cannot be used over butyrate. The main difference between the two dopes is the film base. Nitrate uses a special cotton dissolved in nitric acid, while the cellulose fibers in butyrate dope have been dissolved in acetic acid and mixed with butyl alcohols. The plasticizers in the two dopes are different and the resin balance and solvent balances are different.

It can actually be said that there are two reasons for poor adhesion of the dope in a polyester fabric: poor operator techniques, in which the prime coat does not encapsulate the fibers — this can be from the dope being too thick or by it not being pressed into the fabric. The second reason for a dope finish separating is too much aluminum powder in the finish. Three-and-one-half ounces by weight of paste per gallon of unthinned dope is the absolute maximum to use. After the aluminum dope is sanded, the surface must be scrubbed with water and wiped clean of any loose aluminum powder.

3. Pinholes And Bubbles

Aircraft dope contains between 8% and 45% solids, while the rest is solvents which will disappear — evaporate. This amount of liquid must change into a gas. The tiny bubbles join together in clusters and rise to the surface of the dope as a rather large bubble. When the bubble breaks, it forms a crater. This is illustrated in Figure 19.

Air entrapped into the dope forms the same type of pinholes as does the solvent evaporation. This air may be introduced into the dope by too fast agitation. The pressure pot should have an air-driven agitator which turns about 20 to 50 RPM, no faster than you can turn it by hand.

Excessive atomizing air will cause tiny bubbles to form in the dope film. If you look across a surface which has been sprayed with clear dope using too much atomizing air pressure, the surface will have little sparkles where the light strikes the tiny dope bubbles which have come to the surface and not broken. There should be no more air pressure used than necessary to properly atomize the dope.

Air embedded in the reinforcing tape under the surface tape will attract solvents from the dope. When the second coat of dope is applied, bubbles will form under the tape, Figure 20. To prevent these bubbles, saturate the reinforcing tape with clear dope before it is laid over the ribs. After the surface tape is put on, work it down to the surface, forcing out all of the entrapped air.

Pinholes in the finish may also be caused by case hardening of the surface material. If air blows over the surface and dries the top into a film which will not allow the gases in the still liquid material to escape, these gases will join together and build up into fairly large bubbles. When they get large enough to break through the surface film, they leave a rather large pinhole with cratered edges. This can be prevented by using multiple thin, wet coats of dope rather than fewer, heavier coats.

4. Blushing

Blushing is probably the most common trouble with dope. It is the white or grayish cast that forms on a doped or lacquered surface. If the humidity is too high, or if the solvents evaporate excessively fast, the temperature of the surface drops below the dew point of the air, and moisture condenses on the surface. This water causes the nitrocellulose to precipitate or drop out. If there are not enough solvents left in the dope to redissolve this nitrocellulose, it will form on the surface as blush.

Dope is a rather complex chemical, consisting basically of the film base and several solvents. These different solvents are "stepped," meaning that some of them will evaporate almost as soon as they leave the gun; others will reach the surface, then evaporate; while others will dry even more slowly.

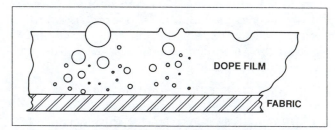

Figure 19. As the solvents evaporate from the dope, they form tiny bubbles. As these bubbles rise to the surface, they join others to form larger bubbles. When they reach the surface and break, they form craters, or pinholes.

Each of these have their purpose in developing the desirable film.

Blushing can usually be controlled or removed by spraying a coat of dope over the blushed area which has had some of its thinner replaced by the much slower drying retarder. This is called a wash or "burn down" coat. The solvents attack the surface and re-flow it. Usually if the blush is not too bad, the next coat of dope will flow it out smooth. Spotty blushing after the last coat of dope may be burned out by spraying it with a mixture of one part retarder and two parts thinner sprayed on in a *very light mist coat*. Don't wet the surface. Wait ten or fifteen minutes and lightly mist it again. This will work with light blush, but if it is too severe, the surface will have to be sanded to remove all of the blush and resprayed when the atmospheric conditions are more suitable.

5. Fisheyes

Fisheyes is a condition of the finish in which there are isolated patches, or areas which have not dried. This is usually caused by some oil, wax, or silicone product which has reached the surface. This contaminant rejects the finish, leaving a bare spot upon the surface.

One unique condition has shown up, especially with butyrate dope finishes: the formation of fisheyes for no apparent reason. This condition has shown up as imperfections in freshly painted airplanes in widely separated geographical locations. Some detective work showed that in each instance, the shop doing the finishing work was under a Standard Instrument Departure (SID) route

for a major jet airport. It seems apparent that the jets, on departure, dumped the fuel which had collected from the nozzle manifold drain on shut-down. This small amount of turbine fuel appears to cause enough contamination in the air to show up as fisheyes in a butyrate finish. There is no reason for this to be unique to butyrate finishes, but they are the ones which have been most affected.

Fisheyes, since they are caused by surface contamination, can be eliminated by carefully scrubbing the surface with toluol before spraying the finish.

6. Dope Roping

Both nitrate and butyrate dopes are viscous, and are quite sensitive to the temperature. If you should attempt to brush on dope which is too thick, or if the temperature is too low, the solvents will flash off while you are still brushing, and the brush will drag across the dope. The dope and the surface should be the same temperature, and slow evaporating solvents should be used.

The technique for applying dope is different from that used for painting a house. The brush should be filled with dope, put on the surface, and stroked across, then lifted off. Don't work the brush back and forth in the dope; one pass is enough. The only purpose of the brush is to carry the dope to the work.

Camel's hair brushes are the best for dope application, but they are quite expensive. Nylon brushes work acceptably well if the size and bristle length are suited for the work. Nylon doesn't hold or release the dope quite as well, however, as the more expensive brushes.

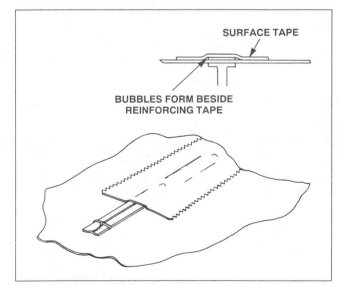

Figure 20. Bubbles will form beside the reinforcing tape under the surface tape if the reinforcing tape is not saturated with dope before it is put on.

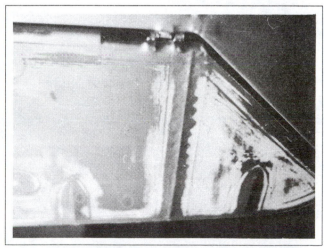

Figure 21. A spotty blush condition may be corrected by spraying a very light mist coat of retarder and thinner on the blush. The slow drying solvbents will re-flow the blush if it is not too severe.

QUESTIONS

1. Does filiform corrosion form on a metal surface because of an improperly cured primer or an improperly cured topcoat?

2. What two problems can cause poor color matching?

3. What color base coat is used to get a proper color match when using acrylic lacquer?

4. What three variables determine the color match of a metallic material?

5. What is applied over a vivid color coating to prevent its fading?

6. What can be done to minimize air impingement haze when spraying acrylic lacquer?

7. Will polyurethane enamel cure more rapidly when the air is warm, or when it is cold?

8. Which dope has a slower burning rate, nitrate or butyrate?

9. What is the main advantage of nitrate dope over butyrate as a prime coat?

10. What two major reasons are there for finish deterioration on polyester?

11. How can you prevent bubbles from forming under surface tape which is applied over reinforcing tape and rib lacing?

12. What is the white or grayish material that forms on a doped surface when it blushes?

13. How does a retarder prevent or minimize blushing?

14. What are fisheyes?

15. Name two causes of dope roping.

CHAPTER V

Special Finishes And Finishing Products

A. High-Visibility Finishes

The need to make aircraft more visible, both on the ground and in the air, has caused the paint manufacturers to develop a series of vivid color finishes for airplanes. Normally these finishes are not used for the complete airplane, but are used for wing-tips, cowlings, the empennage, or colored bands around the fuselage.

These finishes consist of a coat of transparent pigment applied over a white, reflective base coat. Over the colored pigments is sprayed a clear, ultraviolet absorbing topcoat, to help retard fading of the vivid transparent pigments. If the airplane has a good white finish on it, this may be used for the base coat.

Light penetrates the transparent topcoats and it is reflected off of the base; we see the colored reflection.

The application of these finishes is the same as for any sprayed-on finish. Be sure the reducer, the pigmented material, and the topcoat material are all compatible. Paint manufacturers sell all of this material in kit form, so you will have enough of all of the components for the relatively small amount used for the typical application. It is also available in bulk for larger applications.

B. Wrinkle Finish

Instrument panels, electronic equipment, and other aircraft parts subject to considerable rough treatment may be finished with a wrinkled surface. This is essentially a material with very fast drying oils. The surface dries first and as the bottom dries, it shrinks, pulling the surface into the desired wrinkle pattern. The size of the wrinkles is determined to a great extent by the formulation of the material.

All wrinkle-finish material should be sprayed on the surface with a heavy coat and allowed to dry in the way recommended by the manufacturer. There are two types of finish: that which cures by heat, and must be baked to produce the proper wrinkle, and that which dries in the air and does not require baking.

Air-dry wrinkle finish available in aerosol spray cans is considerably softer than the baked finish and is not recommended where there will be much handling or wear. It is also not recommended that air-dry finishes

be baked because the pigments will discolor, and you will not end up with the color you want.

C. Flat Black Lacquer

A durable, non-reflective coating for instrument panels and glare shields is flat black lacquer. This is sprayed on, either with a gun, or from an aerosol spray can, and allowed to air dry. Flat finishes must be put on thin so they will not flow out and gloss. If they are put on too thick, there will be spotty areas of glossy finish in the predominantly flat coat.

D. Wing Walk Compound

A special sharp grain sand is mixed into a tough enamel material to form non- slippery surfaces on airplanes for wing walks or any part of the airplane where a rough surface is desirable. It may be either brushed on with a coarse brush, or sprayed, using a special nozzle for the dense, highly abrasive material. It may be applied directly over the regular finish after thoroughly cleaning the surface, and breaking the glaze if the finish is old. Wing walk compound may be thinned with toluol, or other conventional enamel thinners.

E. Acid-Proof Paint

Battery boxes are one of the more corrosion-prone areas in an airplane because of the continual presence of acid fumes and occasional spilled acid. To prevent damage to the metal, battery boxes and all of the area surrounding the battery should be treated with an acid-proof paint. One of the commonly used materials is a black asphaltum material which resembles tar. It is thinned with toluol and brushed onto the surface after every trace of corrosion has been removed and the corroded area treated with a conversion coating such as Alodine.

An acid-proof finish, far superior to the black asphaltum paint, is a good coat of polyurethane enamel. When an airplane is being painted with polyurethane, a good coat on the battery box and the adjacent area will provide protection from the fumes of the lead-acid or nickel-cadmium batteries, and will not chip or break away from the metal. Polyurethane will not wash away with gasoline or any ordinary solvents.

F. Float Bottom Compound

Seaplane floats take a beating, both from the abrasion of the water and from rocks on the beach or floating debris. In addition to the mechanical damage, they are subject to the maximum exposure to corrosive elements. For protection to the bottom of the float, a material similar to acid-proof paint may be used. It is an asphaltum product and is thinned to spraying consistency with toluol. If a black finish is not desirable on the bottom of the float, aluminum paste may be suspended in toluol and sprayed onto the black compound; the toluol will soften the material and allow the aluminum powder to embed in the finish.

Polyurethane enamel will provide a good abrasion-resistant finish for floats and may be used instead of the more conventional float bottom compound.

G. Fuel Tank Sealer

Built-up fuel tanks may develop seep leaks around rivets and seams. These can be stopped with a resilient, non-hardening tank sealer. Tanks which can be removed from the structure may be sealed by sloshing them with the sealer.

1. Tank Preparation

a. Drain the tank and ventilate it thoroughly.

b. Remove the tank from the airplane.

c. Remove the gage sender, vent line fittings, main line screen, and quick drain.

d. Rinse the tank with clean white gasoline to get rid of all of the fuel dyes, and air-dry the tank for at least thirty minutes at room temperature. If the tank has been previously sealed, remove all of the old sealer by pouring about a gallon of acetone or ethyl acetate into the tank and sealing it up for an hour or two. The vapors will soften the sealer, and the liquid may then be sloshed around in the tank and dumped out. Repeat the process until the solvents come out clean and an inspection shows that there is none of the old sealant in the tank. Drain it completely and dry the tank with compressed air.

e. Plug all the threaded holes with pipe plugs and cover the gage hole with tape or a metal plate.

f. Pour about a gallon of sloshing or sealing compound into the tank, thinned as recommended on the can. Cover the filler hole and slowly rotate the tank until every bit of the inside is covered. Leave the main line plug slightly loose to relieve pressure which builds up during sloshing.

g. Place the tank over a container and remove the quick-drain plug. Allow as much compound to drain out as will. This compound will remain usable if it is covered immediately after draining it from the tank. If it has thickened, it can be thinned with ethyl acetate or methyl-ethyl-ketone and mixed thoroughly.

h. Reinstall the drain plug and put on another coat by pouring a gallon of sealer in the tank and rotating the tank as before.

i. Drain and dry the tank for at least 24 hours; or if the air, under a pressure not exceeding 1½ psi, is circulated through the tank, it can be used after 16 hours.

j. Clean all of the threaded openings with a bottle brush and MEK or ethyl acetate, and install the fittings, using an appropriate thread lubricant.

k. Coat the float of the sender unit with vaseline or light grease to prevent its sticking to the compound, and reinstall the sender, using a new gasket and Permatex No. 2.

l. Reinstall the tank according to the manufacturer's recommendations and fill it with fuel. Check the operation of the sender, and free it from the bottom of the tank if it has stuck in the fresh sealer.

Never reseal a tank until every bit of the old material has been removed. No new material may be put in the tank if even a trace of the old material remains.

This type of sealant may be used around rivets and seams in built-up tanks if the area is perfectly clean and scrubbed with ethyl acetate or MEK and the sealant brushed into the seams and around the rivets inside the tank. Brush it onto a thickness about the same as you would have from the two coats of sloshed sealant. Cure the sealant by flowing air through the tank for 12 to 24 hours. Use just enough pressure to keep air moving through the tank.

H. Seam Paste

This is a thick zinc chromate material with asbestos fibers embedded in it. It is used for making waterproof joints in seaplane hulls or in floats and to make leakproof seams in fuel tanks. It is also used as a dielectric for joining dissimilar metals. It is put on with a putty knife or squeegee and smoothed down to the desired thickness and the seam joined. This material will not harden.

I. High Temperature Finishes

1. Engine Enamel

This enamel has pigments that are colorfast under high temperatures. The special colors used by engine manufacturers are available in this material. It is thinned with regular enamel reducer or toluol.

2. Heat-resistant Aluminum Paint

This material is especially designed to resist temperatures up to about 1200 degrees Fahrenheit. It is used on exhaust systems and heater shrouds.

J. Rot-Inhibiting Sealer

Organic materials such as aircraft woods are subject to fungus or mildew which destroys the strength of the fibers. A special alkyd resin having very low solids content is mixed with fungicidal materials, and wood structures may be treated with this before they are varnished. Large and intricate wood structures may be dipped into a vat of this material to be sure every portion is protected. This should dry at least 24 hours before covering it with spar varnish.

Rot-inhibiting sealer, as any fungicide or mildew-cide, is poisonous, but because of its extremely low toxicity, no special safety precautions are required for its use, beyond adequate ventilation.

K. Spar Varnish

Spar varnish is a phenolic modified oil which cures by oxidation rather than evaporation of its solvents. It produces a tough, highly water-resistant film which is not softened by the solvents used in the varnish. It is used over the rot-inhibiting sealer for aircraft wood structures.

Electronic components such as circuit boards are often protected by a topcoat of spar varnish in which a fungicide is dissolved. This is a transparent coating having a slight amber cast, and it must be completely removed from any point to be soldered.

L. Tube Oil

A thin, non-hardening, raw linseed oil is used to protect the inside of tubular structure in aircraft fuselages, empennage structure, and landing gear. A hole is drilled into each tube section and tube oil forced in. The structure is rotated so the oil will fill every portion of the tube and then the oil is drained out. After the oil has drained, the holes are plugged with sheet metal screws or by welding.

M. Thinners And Reducers

Dopes, enamels, and lacquers are formulated in such a way that the pigments or film material is suspended in the appropriate solvents. These may be thinned or reduced to make them less viscous for spraying. Addition of the correct type and amount of thinner is of the utmost importance. Be sure to use only the thinner or reducer recommended by the paint manufacturer, and thin to the proper viscosity by mixing the proper parts of material and thinner as specified; or better, by the viscosity called for or known to be correct from experience. This may be tested by a viscosity cup. More about the use of this cup is included in the section on paint shop equipment.

1. Nitrate Dope Thinner

Nitrate dope thinner, some of which meets Federal Specifications TT-T-266C, may be used to thin nitrate dope, nitrocellulose lacquers, or nitrate cement. This thinner, if rubbed on a dry dope film, will determine whether the dope is nitrate or butyrate. If the film softens immediately, it is nitrate; if it does not, it is butyrate.

2. Butyrate Dope Thinner

Butyrate dope thinners can be used in butyrate or nitrate dope, but nitrate thinners cannot be used in butyrate. Acrylic lacquer thinner may be used in either butyrate or nitrate dope, but its use is generally not recommended. Butyrate thinner cannot be used to thin acrylics. There is a universal thinner that will thin nitrate, butyrate, or acrylic lacquers, but because of the special requirements of each of these products, this type of material is a compromise and is not generally recommended. It is always best to use thinners and reducers made specifically for the product you are thinning.

3. Retarder

Retarder is a special type of thinner having rich solvents. These dry very slowly and prevent the temperature drop which condenses moisture and causes blushing. If dope spraying must be done in times of high humidity, and there is no way to control the amount of moisture in the air, retarder may be used in place of some of the regular thinner. One part of retarder to four or five parts of regular thinner is about the most that will do any good. One part of the retarder to three parts of thinner is the absolute maximum that will do any good.

A mixture of one part of retarder and two parts thinner, very lightly mist-coated over a blushed surface will sometimes remove blush. There is more about this in the section on blushing.

4. Anti-blush Thinner

Show-job-type fabric finishes consist of many coats of dope, sanded with number 600-grit paper between each coat, and rubbed down after the last coat. These coats are sprayed on wet and thin. Anti-blush thinner is used in this type of finish because its slower drying solvents allow each coat more time to flow out and form a smoother film. Anti-blush thinner is between regular thinner and retarder in its solvents and drying time.

5. Enamel Reducer

There are several proprietary reducers on the market for enamel, but a good reducer such as toluol does a satisfactory job reducing enamels, engine enamels, wing walk compounds, zinc chromate primer, acid-proof black paint, float bottom compound, and white, dope-proof paint. In addition to using it as a reducer for these materials, it is in most cases satisfactory to use to wash down a surface to remove the wax after a paint stripper has been used.

6. Methyl-ethyl-ketone (MEK)

This is one of the most universally used solvents for general cleaning. It is used to remove all wax and grease from a surface prior to spraying a finish, and it finds use in the engine shop around the ignition and carburetion systems for cleaning and decarbonizing parts.

7. Acetone

Like MEK, acetone is quite universally used as a solvent and cleaner. It is used to remove lacquer finishes and for clean-up after painting. It will soften acrylics or lacquers that have set-up in spray guns or hoses. But, it has little effect on polyurethane which has cured in the gun or lines. When buying acetone, be sure to get only virgin acetone, as recovered acetone is often so acidic that it can damage anything on which it is used.

N. Rejuvenator

Aircraft dope consists of the film base, solvents to dissolve the base, and plasticizers to make the base flexible and resilient. Exposure to time and sunlight cause the film to lose its resilience and become brittle. When this happens, the dope will crack or ringworm and open the fabric to the harmful sun's rays. If the fabric is still good as proven by either a punch test or a pull test, the dope may be rejuvenated.

Rejuvenator is composed essentially of potent solvents and a plasticizer. Tri-cresyl-phosphate (TCP) is a permanent, somewhat fire-retardant plasticizer used in many rejuvenators. The weathered surface should be washed and watersanded to remove any old wax or polish and a good heavy coat of rejuvenator sprayed on the surface. This softens the old dope and flows the cracks back together. After the first coat, the fabric will loosen, but a second coat will restore the original tightness. Sand the surface and spray on a good cross-coat of aluminum butyrate dope. After wet-sanding this, thoroughly remove all of the traces of aluminum sanding dust by scrubbing down with water, wiping clean and allowing to dry. Then, you may spray on a finish coat of colored butyrate dope.

If a fabric job has been interrupted in the process of finishing, and the aluminum dope has been on the fabric for a considerable time before the topcoats are applied, spray on a coat of rejuvenator to soften the aluminum dope, and then spray on the color coats.

O. Spot Putty And Sanding Surfacer

Nitrocellulose spot putty is used to fill cracks or low spots in wood skins before they are covered with fabric. Be careful when using spot putty that defects which may cause loss of structural strength are not covered up and hidden. If the skin is to be covered only with a film of enamel, enamel spot putty should be used so the solvents in the putty will not lift the film.

Sanding surfacer is applied over a wood or fiberglass laid-up structure to fill the surface irregularities with a material having enough body that it can be sanded smooth. The material used in automobile body shops is satisfactory for use on these parts.

QUESTIONS

1. What is the purpose of the transparent top coat applied over a vivid-color, high-visibility finish?

2. Why should air-dry enamel or wrinkle finishes not be baked?

3. Should flat black instrument panel lacquer be sprayed on thick or thin?

4. What is used to thin acid-proof black paint?

5. What can be used to remove all of the old fuel tank sealant before applying new sealer?

6. What is the main difference between engine enamel and conventional enamel?

7. Can nitrate dope thinner be used to thin butyrate dope?

8. What is rejuvenator?

CHAPTER VI

Finishing Equipment

A. Paint Storage

The flammable nature of aircraft finishing materials requires special handling for them. The storage facilities must comply with Occupational Safety and Health Act (OSHA) requirements regarding air circulation, lighting, and fire protection.

The storage areas should have sufficient heat to prevent the material freezing, if possible. Freezing will not actually damage many of the materials, but they require a long time to warm to proper application temperature. The airplane, the room, and the material should all be at the same temperature for the best adhesion.

There should be a retaining curb around the storage area to contain any material that may leak. If a drum of dope or thinner should break it should be contained near the drum rather than spreading all over the storage room. If the facility is in a warm part of the country, an outside storage room is a good consideration; it should be roofed over, curbed, and ventilated top and bottom to prevent heat buildup from the sun. It should be locked to prevent unauthorized people or children getting to the materials. All of the stock should be rotated so that all material bought first will be used first. Pigmented materials such as zinc chromate and polyurethane enamels should be stored with the cans inverted. At each inventory check, they should be turned over so the pigments will not have so much opportunity to pack on the bottom of the can.

All containers of flammable material, whether full or empty, constitute a fire hazard and should be properly stored and cared for. When empty, they should be disposed of properly.

B. Spray Booth Or Spray Area

An aircraft maintenance shop specializing in painting will normally have a special paint hangar with temperature and humidity control and provisions for cleaning and circulating the air which flows through the booth.

Any shop anticipating building a paint hangar should seriously consider the extent of the business anticipated, so that the shop can be designed adequate in size, neither too large nor too small. A properly designed and built paint shop is costly, and

if it is not used sufficiently to get a good return on the investment, it may be looked on with disfavor by the management or the stockholders.

After the decision has been made to build the facility and the space allotted for it, some of the major paint finishing equipment manufacturers such as DeVilbiss or Binks should be consulted for recommendations on such items as the exhaust and air make-up systems and temperature and humidity controls. All of the air exchange system must meet the requirements of OSHA. Recirculating the air is feasible only if there are adequate means of scrubbing it of all solvents and solids.

All electrical switches and outlets must meet OSHA specifications, and all lights must be explosion-proof. Smooth concrete floors are usually acceptable, but it is best to consult with your insurance company for their ruling on the matter of floors and other questionable areas.

Many airplanes, however, are painted in facilities far less than the optimum — perhaps in the corner of a hangar, modified into a spray room by enclosing it with polyethylene sheeting and removing the fumes with an exhaust fan.

There must be sufficient movement of the air when spraying, so there is no more than a slight odor of the finishing material. The fan should be near the floor and be belt-driven, with the motor located away from the fumes. While very little aircraft finishing material is toxic, it is not advisable to breathe their fumes, because they deplete the oxygen supply required by the body.

C. Air Compressors, Storage, And Distribution Lines

One of the most important pieces of equipment used for aircraft finishing is an air compressor adequate for the job. It should supply enough clean, dry air to carry all of the spray guns you anticipate using at any one time, with the proper pressure *AT THE GUN*. A rule of thumb for air compressor size is one horsepower on the compressor will produce three to five cubic feet of air per minute, at 50 psi.

It is poor economy to try to save money by buying an air compressor smaller than actually needed. When anticipating this purchase, show the compressor salesman exactly what your maximum needs are likely to be, and allow for a growth factor. If you try to economize here, you are likely to end up with a compressor inadequate for the job. The manufacturers of spray equipment are in a position to advise on the size and type of compressor, filters, water traps, and transformers you will need for your specific application.

The air receiver, or storage tank, should have a water drain trap which can be drained every day and should have a pop-off safety valve which, in case of compressor cut-out switch failure, will prevent the air pressure in the receiver becoming excessive.

The piping between the air receiver and the spray guns should be large enough so there will be no excessive pressure drop because of the airflow. The pipe should be laid in such a way that any water which condenses in the line will drain back into the receiver rather than flow to the air transformer. Large lines with low air velocity allow large amounts of water to drain back and not be blown on or into the finish.

There must be adequate water drain traps in the lines of the compressed air system, and these traps should be drained on a regular basis. The drain trap and filter in the air transformer should not be depended upon for the main filtering. However, this should be drained every morning; more often if the air is specially humid. The filters in the line should remove any oil from the compressor as well as all of the water which collects in the lines.

The pressure on the line at the transformer is not necessarily the pressure at the gun, and if this drop is not considered, it is easy to end up with far too little pressure for atomization. If there is any doubt as to the pressure drop in your line, a simple check with an accurate air pressure gage screwed into a T-fitting at the gun inlet will allow you to make a chart comparing the air transformer pressure with the pressure actually at the gun.

D. Spray Equipment

1. Air-atomized Spray Equipment

There are several types of spray guns on the market, but the most common are the bleeder-type gun, the pressure-fed gun, and the syphon or suction cup gun. Bleeder-type guns are used for low-pressure applications, and actually find no place in the professional aircraft finishing shop. Pressure-fed guns are the most popular for spraying a complete airplane, and are found in almost every shop where this type work is done. A five gallon pressure pot with an air-driven agitator is adequate for most jobs. A suction cup gun is necessary for spraying the numbers, the trim, and for touch-up work. While these guns are essentially the same, they should be kept separate, and the suction cup gun not used on the pressure pot. In the discussion of equipment, only the pressure-fed and the suction cup guns will be discussed.

Figure 22. *The air transformer and drain trap provide water-free air of the correct pressure for operating the spray gun. The water should be drained each morning, or more often if the air is extra humid.*

Figure 23. *A typical two stage air compressor with air receiver. The air supply should carry all of the guns you antaicipate using at one time, with adequate pressure AT THE GUN.*

40

In aircraft painting, as with every other aspect of aviation maintenance, it is important that only quality equipment be used. Trying to save money by buying less than the best spray gun is false economy, as it will result in less than satisfactory finishing. DeVilbiss and Binks both make guns of high quality, guns which may be fitted with an assortment of fluid tips and needles to match the gun to the material being sprayed.

With the pressure-fed gun, the material is put into a pressure pot and air pressure is used to force the material to the gun. The amount of pressure on the pot determines the amount of material which will be deposited on the surface being sprayed. The atomizing air is fed directly to the gun and is controlled by a regulator, independent of that used on the pot. Two- and five- gallon pots are commonly used in aircraft shops. These pots can have either air-driven agitators or hand-turned paddles inside the pot to keep the pigments in suspension in the material.

Syphon-fed, or suction cup guns are used for smaller amounts of spraying, such as trim, for registration numbers, or for component painting. The material is held in a quart cup and drawn into the atomizing airstream by a suction created by this airflow.

2. Airless Spray Equipment

While used almost universally for production work, airless spray equipment does not find widespread use in maintenance shops. In this type of spray equipment, the material is pumped under high pressure (from 500 to 4500 psi) to a small orifice in the

Figure 24. Stepped down pressure on the pressure cup forces the correct amount of material to the gun.

nozzle where it is released. Releasing this high-pressure fluid breaks it down into tiny droplets with enough momentum to carry them to the surface being sprayed. Airless equipment deposits about the same type of film on the surface as an air-atomized system, but the airless equipment does not have nearly so much overspray. About the same amount of solvents are required with either type of equipment, but those used in the airless spray must be slower drying.

A normal, low-pressure air compressor drives an air motor which operates a reciprocating airless fluid pump. This pump directs the high-pressure fluid to the gun where it is discharged as a finely atomized spray. Caution should be used with airless equipment, as the stream of high-pressure material, especially with the nozzle removed from the gun, can inflict injury if it should hit a person.

3. Electrostatic Spray Equipment

Another production type of spray equipment that finds effective applications in factories but limited use in the field is electrostatic spray equipment. In this method of application, a high-voltage difference is set up between the material and the work. When the paint is sprayed, the electrostatic charge attracts the droplets of paint, and they wrap around the material and coat even the side away from the gun. The material may be atomized by air, by an airless pump, or by special electrostatic atomization systems using a spinning disc.

E. Respirators And Masks

Paint-spraying operations should not be conducted in areas where the air is highly contaminated. Naturally there will be some solvents and solids contamination from the spraying, but a good filter-type mask will remove the solids, and the airflow in the paint room should be adequate to remove all of the solvent fumes.

An airflow-type mask is desirable for prolonged spraying, but care must be taken that only good clean air be fed into the mask.

F. Measuring Equipment

1. Graduates

Most paint mixing is done on the basis of so many parts of material to so many parts of thinner. In order to do this accurately, there should be some graduates in the shop. You will need at least a 16-ounce (one pint), a 32-ounce (quart), and a one-gallon measure. The larger one is quite handy for measuring and mixing fungicidal dope and aluminum-pigmented dope. The best graduates are made of stainless steel, and these will last for a lifetime. Linear polyethylene graduates are available in almost all sizes and are less expensive than stainless steel. Graduates made of this material are not affected by the commonly used solvents and may be cleaned by washing them in thinner; or, if the paint is allowed to dry in them, the film will peel out without leaving any residue.

2. Viscosity Measurement

Many experienced painters judge the viscosity of a material by watching it drip off of the stirring paddle. With years of experience, this gives good results, but for those with less experience to draw on, and for materials which are new to you, a more precise system is advantageous. Paint laboratories use either a Zahn or a Ford cup to measure viscosity. These are metal cups of specific size and shape, having a small hole in their bottom. The cup is filled completely full of material, and the viscosity of the material is determined by the time, in seconds, required for the cup to empty through the hole to the point that there is a break in the flow.

Figure 25. Respirator masks are used to remove solids from the air breathed by the painter. Adequate ventilatiaon should be used to remove toxic fumes from the air.

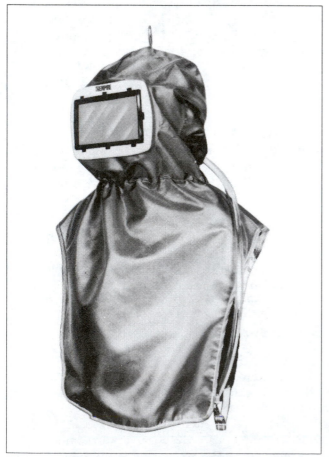

Figure 26. Airflow type respirators should be used if the concentration of fumes is excessive.

Figure 27. The viscosity of a material is measured by the number of seconds required for this cup to empty to the point of THE FIRST BREAK in the flow through the hole in its bottom.

Both the Zahn and the Ford cups are quite expensive and are more for use in the laboratory than in an aircraft paint shop. A very satisfactory viscosimeter is sold by Sears and costs only a couple of dollars; Figure 27. Some of these are made of metal, and others are made of polyethylene plastic. To use this cup, mix the paint to the spraying consistency you like best, then take the cup and dip it in the material. Using a watch with a sweep-second hand, or better, a stopwatch, time the flow of material from the time you lift the cup from the container until the *first break* in the flow occurs. There will be more flow after this first break, but disregard it. The time is taken only to the first break. If the viscosity of a satisfactory batch of paint is checked, it is a simple matter to mix the material and thinner to exactly the same viscosity each time you are going to spray. Be sure to keep the cup perfectly clean, and do not allow it to get scratched or bent. It is a good idea to keep it hanging up when it is not in use.

G. Mixing Equipment

Shops which do a considerable amount of painting find a mechanical paint shaker which holds a five-gallon can a most convenient piece of equipment. The can of pigmented material is put on the shaker upside down and agitated for fifteen to twenty minutes. This will assure that every bit of the pigment is mixed into the vehicle and is then ready to thin.

Small shops which do not have a shaker can do a satisfactory job of mixing, though paying a penalty in the time involved, by following the procedure outlined in the section on color matching — Chapter 4.

Hand-held agitators, if driven by an air drill motor, will get some of the harder packed pigments loose, but this should be followed by "boxing" the material. If you use a hand-held agitator, *do not use an electric drill motor*. The agitation stirs up flammable fumes which rise around the sparking brushes in the electric motor and can easily start a flash fire.

If you should get a container fire, put the lid back on the container and smother the fire. *Do not attempt to carry the burning can out of the paint room.* Any kind of cover which will exclude the air from the surface will put the fire out without spreading it.

QUESTIONS:

1. *When adjusting pressure for spraying, which is important, the pressure at the air transformer or the pressure at the gun?*

2. *hat determines the amount of material a pressure-pot gun will deposit on the surface?*

3. *What atomizes the material sprayed by an airless spray gun?*

4. *Should a respirator mask be depended on to remove solvent fumes from the air the painter breathes?*

5. *What is the principle of measuring viscosity with a viscosity cup?*

6. *Why should an electric drill motor not be used or mixing paint?*

APPENDIX A

REDUCTION RATIOS

There are several ways to state the amount a material should be reduced or thinned. Throughout this book, reference has been made to reducing by parts. The following table relates reduction by percentage to reduction by parts.

Percent Reduction	Parts Reduction	
%	Material	Thinner
5	20	1
10	10	1
20	5	1
25	4	1
30	3.3	1
50	2	1
100	1	1

APPENDIX B

DIGEST OF FINISHING PROBLEMS

ROUGH FINISH

Dope too cold or viscous. The aircraft, dope, and thinner should all be at the same temperature, about 70°F, before spraying.

FABRIC WILL NOT TAUTEN

Fabric put on too loose.

Fabric allowed to remain undoped too long.

Too much retarder used for thinning dope.

BLUSHING

Humidity too high.

Moisture in spray system.

Dope applied over a moist surface.

PINHOLING OR BLISTERS

Water or oil in spray system or on surface.

Undercoat not thoroughly dry. Too fast surface drying.

Film coat too heavy.

BUBBLES AND BRIDGING

Dope too cold, or not brushed out properly.

Temperature of dope room too high.

RUNS AND SAGS

Use of improper equipment.

Improper adjustment of equipment.

Improper thinning or faulty spray technique.

DOPE WILL NOT DRY

Oil, grease, or wax on surface.

DULL SPOTS

Porous spot putty or undercoating, allowing dope or lacquer to sink in.

BLEEDING

Organic pigments or dyes used in the under coats which are soluble in the topcoat solvents.

WHITE SPOTS

Water in spray system or on surface.

PAINT OR PRIMER PEELING

Wax from stripper or detergent from cleaning process may still be on surface.

BROWN SPOTS

Oil in spray system.

ORANGE PEEL

Spraying with too high pressure.

Use of too fast drying thinner.

Cold, damp draft over surface.

WRINKLING

Reaction between solvents and primer or undercoats.

OVERSPRAY

Wrong spray technique.

Too fast drying thinner.

FISHEYES

Silicone, wax, or polish contamination on surface.

Oil from air compressor.

APPENDIX C

MATERIAL REQUIREMENTS FOR REFINISHING TYPICAL LIGHT AIRPLANES

FABRIC-COVERED AIRCRAFT SUCH AS TRI-PACER

Nitrate cement, for fabric attachment . . . . 1 gallon

Nitrate clear dope (for first coat) . . . . 10 gallons

Nitrate Thinner . . . . 10 gallons

Butyrate clear dope . . . . 25 gallons

Butyrate thinner . . . . 25 gallons

Colored butyrate dope (basic color) . . . . 10 gallons

Colored butyrate dope (trim color) . . . . 1 gallon

Butyrate retarder . . . . as required by conditions

Aluminum paste . . . . 2 pounds

Epoxy primer and mixing liquid. . . . . 1-1/2 gallons ea.

Enamel color (for metal) . . . . 1 quart per color

Enamel reducer . . . . 1 quart

NOTE: For finishing an aircraft such as a Bellanca Viking, double the quantity of clear dope and thinner.

AVERAGE SINGLE-ENGINE, 4-PLACE ALL-METAL AIRPLANE

ACRYLIC LACQUER SYSTEM

Wash Primer System

    Wash primer . . . . 4 quarts

    Acid diluent . . . . 2 pints

    Thinner . . . . 4 quarts

Epoxy Primer System (alternate and preferred method, if time permits).

    Epoxy primer . . . . 4 quarts

    Epoxy catalyst . . . . 4 quarts

    Epoxy primer reducer . . . . 2 gallons

    Acrylic color . . . . 6 gallons

    Acrylic thinner . . . . 10 gallons

POLYURETHANE SYSTEM

    Epoxy primer . . . . 4 quarts

    Epoxy catalyst . . . . 4 quarts

    Epoxy primer reducer . . . . 2 gallons

    Polyurethane color . . . . 2 gallons

    Polyurethane catalyst . . . . 2 gallons

    Polyurethane reducer . . . . 2 quarts

Refer to the appropriate portion of the text for the application technique and curing time.

APPENDIX D

METRIC CONVERSIONS

BASIC METRIC PREFIXES

Micro	1/1,000,000
Milli	1/1,000
Centi	1/100
Deci	1/10

UNIT

Deka	10
Hecto	100
Kilo	1,000
Mega	1,000,000

LINEAR CONVERSIONS

1 inch = 2.54 centimeters

1 centimeter = 0.3937 inch

VOLUME CONVERSIONS

1 cubic inch = 16.387 cubic centimeters (cc)

1 cubic centimeter = 0.061023 cubic inch

CAPACITY CONVERSIONS

1 gallon = 3.785 liters

1 liter = 1.0567 quarts

APPENDIX E

FAMILY TREE OF AIRCRAFT FINISHING MATERIAL

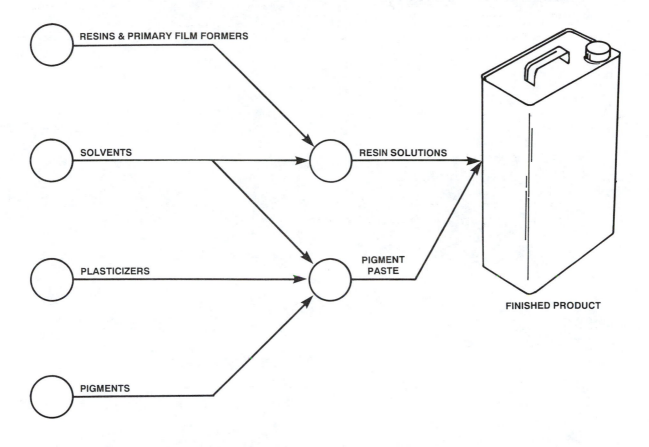

GLOSSARY

This glossary of terms is to give a ready reference to the meaning of some of the words with which you may not be familiar. These definitions may differ from those of standard dictionaries, but are more in line with shop usage.

acetone A flammable liquid ketone (C_3H_6O) used as a solvent and a constituent in many aircraft finishes.

acid diluent A constituent of wash primer that provides a mild etch to the surface of the metal for a better bond.

acrylic lacquer An aircraft finish consisting of an acrylic resin vehicle and certain volatile solvents.

air impingement A fault in a finish resembling a haze. It is caused by microscopic air bubbles carried to the surface in the paint, caused by excessive atomizing air pressure.

alkyd resin A type of synthetic resin used as the base for certain enamels and primers.

alloying agent A metal other than the base metal in an alloy.

anti-blush thinner A slow-drying thinner used in conditions of high humidity to prevent dope blushing.

asphaltum A bituminous or tar-base material.

bleeding reds Certain red pigments which are soluble in the solvents used for their application. They will bleed up through any coat of material put over them.

blushing The white or grayish cast which forms on a lacquer or dope film which has been applied under conditions of too high humidity. It is actually nitrocellulose which has precipitated from the finish.

bond An attachment of one material to another, or of a finish to the metal or fabric.

boxing a paint Procedure in which pigmented paint is thoroughly mixed by pouring it back and forth between two containers.

burn down coat A coat of dope with some of its thinner replaced with retarder, sprayed on a blushed area to attack and re-flow the surface and remove the blush.

catalyst An agent which causes a chemical reaction within the solution.

catalyzed material A material whose cure is initiated by the addition of a catalyst.

Ceconite® A synthetic polyester-filament fabric fiber.

clad aluminum Aluminum alloy which has a coating of pure aluminum rolled onto both sides for corrosion protection.

color wheel A means of visualizing the color which will result when the basic colors are mixed.

contact cement A syrupy adhesive applied to both surfaces which bonds on contact.

conversion coating A chemical solution used to form a dense, non-porous oxide or phosphate film on the surface of aluminum or magnesium alloys.

corrosion A chemical action caused by dissimilar metals acting in the presence of an electrolyte. It shows up as a white powder on the metal surface and actually eats the metal away.

cross-coat A double coat of paint sprayed on in one direction and then before the paint flashes off, at right angles to it.

cure A chemical change which takes place in a finishing system which produces the desired surface.

Dacron® The proprietary name of a synthetic polyester fabric used for covering aircraft structure.

dielectric An electrical insulator.

dissimilar metals Different metals, usually considered to be in contact with each other, which produce an electrical potential difference and are susceptible to corrosion.

dope-proof paint A paint which will not be softened or lifted by dope.

dope roping A condition in the application of dope in which the surface dries while the dope is being brushed. This results in a stringy, uneven surface.

dry-bulb temperature Air temperature without the effect of water evaporation.

electro-chemical Electrical potential difference which exists within a material because of its chemical composition.

enamel A material whose pigments are suspended in a vehicle which cures by conversion of some of its solvents by heat, oxidation, or by catalytic action.

encapsulate   Completely surround each fiber with dope. This provides a bond for all the subsequent coats of dope to the fabric.

epoxy primer   A two-part, catalyzed material used to provide a good bond between a surface and the topcoat.

faying strip   The strip along the edge of a sheet metal skin where a lap joint is formed. This inaccessible area is highly susceptible to the formation of corrosion.

filiform corrosion   A thread, or filament-like, corrosion which forms on aluminum skins beneath polyurethane enamel.

fisheyes   Isolated areas in the finish which have rejected the finish, because of wax or silicone contamination on the surface.

flashing off   Drying of a finish to touch by the evaporation of certain of the solvents. The film is not dry rand hard throughout.

fungicidal paste   Paste which is mixed with clear dope to apply as a first coat on cotton. The fungicidal agent soaks into the fibers and prevents the formation of mold or fungus attack.

fungus spores   Essentially the seed of certain fungi which can enter organic material such as cotton or linen and cause the material to rot.

glaze   The hard, smooth surface of a finishing system. This must normally be "broken" or roughened before another coat of material will adhere to it.

greige Dacron®   A synthetic polyester fiber fabric in its raw or unshrunk condition as it comes from the loom.

hold-out   The ability of a primer to hold the top coats without their sinking into its surface.

hue   The gradation of colors. The characteristic of a color that differentiates between red, blue, or yellow, and any of the intermediate colors.

induction period   A time period after catalyzed material is mixed, in which the material is allowed to begin its cure before it is sprayed onto the surface.

kraft paper   Strong brown paper such as grocery sacks are made of.

metallic pigment   Extremely tiny flakes of metal suspended in paint to produce a metallic-like reflection.

metamerism index   A measurement used for scientific color matching. It indicates the way a pigment will look under varying light conditions.

methylene chloride   A liquid solvent (CH_2Cl_2) used as the active agent in many paint strippers.

methyl-ethyl-ketone (MEK)   An important, low cost solvent similar to acetone. It is used as a cleaning agent to prepare a surface for painting and as a stripper for certain finishes.

mil   A metric term meaning 1/1,000 of an inch. In painting, it usually refers to the thickness of a film.

mildewcide   An additive to dope or sealers used on organic materials to inhibit the growth of mildew.

mist coat   A very light spray coat of thinner or other volatile solvent with little or no color in it.

nap   The short fiber ends which protrude from the surface of a fabric. When the fabric is doped, these fibers become stiff and must be sanded off. This is called "laying the nap."

non-atomizing spray   The application of a material to a surface by a spray gun in which the material is fed in a solid stream rather than in tiny droplets.

paint stripper   A chemical material which penetrates the paint film and loosens the bond to the metal.

phosphate film   A dense, insoluble inorganic film deposited on the surface of a metal which has been treated with a conversion coating.

phosphoric acid etchant   That constituent of a conversion coating which microscopically roughens the surface of metal being treated and deposits a phosphate film.

pigment   A powder or paste mixed with a finish to give the desired color.

pinhole   A defect in a finish which appears as a tiny hole. It is caused by a bubble in the paint film.

plasticizer   A constituent of dope used to provide resilience to the film.

Plexiglass®   A proprietary name for a transparent thermoplastic acrylic material used for aircraft windows and windshields.

polyester   The filament used to make Ceconite® or Dacron® fabric.

polyethylene   A lightweight thermoplastic material that has very good chemical- and moisture-resistant characteristics. It is used for plastic sheeting and containers.

polymerization   A type of chemical reaction in which the material essentially jells.

polyurethane   A polymerized film used as a top coat system for aircraft finishing. It produces a hard, chemical-resistant surface whose slow flow-out time gives it an extra smooth or "wet" look.

porous salt   The type of residue normally left on the surface of a metal which has been attacked by corrosion.

pressure-fed gun   A paint spray gun in which the material is fed to the gun by air pressure on the pot or cup holding the material.

pressure pot   A container holding the material to be sprayed. An agitator keeps the material in motion, and a regulator maintains the proper air pressure on the material to feed it to the gun.

prime coats   The first coats of a finish, used to bond the topcoats to the base material.

primer   A material which provides a sandwich between the topcoats and the metal to provide a good bond.

pull test   A fabric strength test in which a sample strip, one inch wide, is pulled until it breaks.

punch test   A test of the strength of aircraft fabric while on the airplane. A pointed, spring loaded plunger is pushed into the fabric, and the amount of force required to penetrate indicates the strength of the fabric.

rag-wing   A common slang term used to refer to a fabric-covered airplane.

retarder   A slow-drying solvent used to prevent blushing or to provide a more glossy finish by allowing the material a longer flow-out time.

rich solvent   Slow-drying solvent.

sanding coat   A coat of surfacer or heavy-bodied material which is applied and sanded off to fill small surface imperfections and thus provide a smooth surface for subsequent coats.

Skydrol® hydraulic fluid   A synthetic, non-flammable, ester-base hydraulic fluid used in modern high-temperature hydraulic systems.

spectro-photometer   A special device used to determine the way a surface reflects light waves of all frequencies. It is used to analyze paint pigments.

stardust   Air-impingement haze in an acrylic finish.

static electricity   An electrical charge which may be built up on a non-conductive surface by friction.

stepped solvents   Solvents in a finish which have different rates of evaporation. Some evaporate almost instantly; others, more slowly. This provides the desired film.

suction cup gun   A paint gun in which the material is held in a cup attached to the gun and drawn into the atomizing air by suction created by the atomizing airflow.

surface tension   A condition that exists on the surface of a liquid because of molecular attraction; it produces an effect similar to a film stretched over the surface.

tack coat   A very light coat of material sprayed on a surface and allowed to stay until the solvents evaporate; it is then covered with the full wet coat of material.

tack rag   A rag, slightly damp with thinner, used to wipe a surface after it has been sanded to prepare it for the application of the next coat of finish.

thinner   A solvent mixed with a paint material to reduce its viscosity.

toluol   The commercial grade of toluene which is a liquid aromatic hydrocarbon similar to benzene, but less volatile, flammable, or toxic.

tone   A tint or shade of a color. A variation of a hue.

tri-cresyl-phosphate (TCP)   A plasticizer used in rejuvenator to restore resilience to the dope film.

ultraviolet rays   Radiation from the sun with wavelengths shorter than are visible. These rays damage organic materials.

vivid color   One of the highly reflective colors used on aircraft for the maximum in visibility.

wash primer   A sandwiching material which provides a mild etch to the metal surface and a good bond for the topcoats. Wash primer is suitable for high- volume production because of its short curing time.

wet-bulb   Temperature of the air modified by the evaporation of water from a wick surrounding the thermometer bulb.

xylol or xylene   A toxic, flammable, aromatic hydrocarbon, similar to benzene. It is used as a solvent.

Zahn cup   A special cup of definite size and shape, with a hole in its bottom. It is used to measure the viscosity of a material by the number of seconds required for the cup to empty.

ANSWERS TO STUDY QUESTIONS
Aircraft Painting And Finishing

Chapter 1

1. Wax holds the active solvents against the surface until they can penetrate.
2. Paint stripper softens, rather than swells, acrylic lacquer. It must be scraped off.
3. Paint stripper loosens the bond between the primer and the polyurethane film.
4. Forms an insoluble phosphate film on the surface.
5. To allow it to begin its curing action.
6. A very thin film, no more than 0.3 mil (0.0003 inch) thick.
7. After about a half hour.
8. Spray on another coat of primer, this one without the acid.
9. You may add as much as two ounces of distilled water to each gallon of thinner.
10. Only thick enough to slightly color the metal; about half a mil (0.0005 inch).
11. Toluol, or some proprietary thinner.
12. The zinc chromate will lift.
13. Normally about ten parts of enamel to one part thinner. This can be varied widely, however.
14. Generally four parts acrylic material to five parts thinner.
15. These films may flow for three to five days after they are sprayed.
16. Twenty-four hours. Under the most ideal conditions, this may be reduced to about five hours.

Chapter 2

1. The instructions accompanying the Supplemental Type Certificate.
2. There is no way of knowing that the fabric is good unless a pull test is made before the fabric is put on the airplane.
3. Fungus spores can get into the fabric and cause it to rot early.
4. Nitrate dope encapsulates the fibers of the fabric better than butyrate.
5. With the first coat of dope.
6. It forms a light-tight barrier which prevents the ultraviolet rays of the sun damaging the dope or the fabric.
7. Until it is smooth and free of all wrinkles, but not tight enough to damage the structure.
8. Because it encapsulates the inorganic fibers of polyester better than butyrate.
9. Not less than five nor more than forty-eight hours.
10. The weave of fiberglass cloth is so open that brushing would force the dope through the fabric, causing it to run down on the inside.
11.
By pulling the filaments closer together.
12. By sanding or removing it with acetone. Never use prepared paint stripper, as its solvents will attack the resins that bond the fiberglass.
13. Any paint having metallic pigment.

Chapter 3

1. The air pressure at the regulator.
2. The amount of air which flows from the wing port holes.
3. Never more than 10 psi fluid pressure, unless you have an exceptionally long material hose.
4. Four parts color to six or seven parts thinner.
5. Insufficient atomizing air pressure.
6. Too much atomizing air pressure.
7. A buildup of dried material around one side of the fluid nozzle.
8. The hole in one of the wing ports being plugged up.
9. This will flush the passageways and clean the tip of the needle.
10. It will damage the packings.
11. Disassemble the gun and soak the parts in acetone or MEK.
12. You may have to throw the gun away if this happens. Only by digging the material out, can you remove hardened catalyzed material.
13. Six to ten inches.
14. An uneven coat. The side nearest the gun will be heavy, and the side away from the gun will be rough.
15. By all but about two or three inches.
16. This will minimize the problem of overspray settling.

17. Paint the difficult or irregular areas first; then go back and paint the flat areas.

18. Spray over it with a mixture of one part retarder and two parts thinner.

19. The slower drying rate of enamel allows it to sink in rather than dry on the surface.

20. This prevents static electricity causing a spark.

21. They allow static electricity to bleed off of the painter gradually, rather than jumping off as a spark.

22. Flood the floor with water and wet-sweep the overspray.

23. Volatile fumes will rise into the sparking brushes and cause a fire.

24. Near the floor. The solvents are all heavier than air.

Chapter 4

1. An improperly cured primer.

2.   1. Improper mixing.
    2. Improper application.

3. White.

4.   1. Spray pressure.
    2. Amount of thinner used.
    3. Number of coats.

5. A transparent, ultraviolet absorbing topcoat.

6. Be sure the material is thin enough, and keep the atomizing air pressure low.

7. When the air is warm.

8. Butyrate.

9. Nitrate encapsulates the fibers of the fabric better than butyrate.

10.   1. Poor operator technique; lack of penetration of nitrate dope.
    2. Too much aluminum powder.

11. Saturate the reinforcing tape with dope before it is placed over the rib.

12. This is nitrocellulose which has precipitated out of the finish.

13. It slows the drying rate of the dope, preventing the temperature drop which causes water to condense on the surface. It is this water that precipitates the nitrocellulose.

14. Isolated areas where the finish was rejected because of wax, or silicone or other contamination on the surface.

15.   1. The dope was too thick when it was applied.
    2. The temperature was too low for proper brushing.

Chapter 5

1. It prevents ultraviolet rays of the sun fading the finish.

2. These finishes may be discolored by the heat.

3. Spray it on thin.

4. Toluol.

5. Acetone or ethyl acetate.

6. Pigments in engine enamel are colorfast under conditions of high temperature.

7. No.

8. A special mixture of rich, high-potency solvents and plasticizers.

Chapter 6

1. The pressure at the gun, always.

2. The fluid pressure on the pot, and the needle travel at the gun.

3. The material, under high pressure, is atomized when it is released through a small orifice.

4. No; the airflow in the paint room should remove the volatile fumes.

5. It is the time in seconds for a given amount of the fluid to flow through a specific-size hole.

6. The fire hazard caused by the fumes getting into the sparking brushes.

FINAL EXAMINATION
Aircraft Painting And Finishing

Place a circle around the letter for the correct answer in each of the following questions.

1. **Which statement is true regarding the use of paint stripper?**
   A. Paint stripper should be thoroughly brushed into the finish being removed.
   B. Paint stripper will soften a polyurethane film, but will not cause it to wrinkle or lift.
   C. For faster action, apply the stripper and cover it with a polyethylene drop cloth.
   D. Paint stripper causes acrylic lacquers to wrinkle up more quickly than polyurethanes.

2. **A conversion coating:**
   A. is all the primer necessary when finishing aluminum alloys.
   B. forms an organic film on the surface of the metal.
   C. forms an insoluble phosphate film on the surface of the metal.
   D. must never be used on material which has been corroded.

3. **Which statement is true about wash primer?**
   A. Wash primer must never be applied over a conversion coating.
   B. Wash primer may be topcoated after it has cured as little as a half hour.
   C. Wash primer should be sprayed immediately after it is mixed.
   D. Wash primer provides its best corrosion protection if there is almost no water in the air.

4. **Which statement is true about epoxy primer?**
   A. Epoxy primer is used in most high production paint shops because of its rapid cure time.
   B. Epoxy primer is likely to cause filiform corrosion if used under a polyurethane finish.
   C. Epoxy primer is a two component material which provides a good bond between the surface and the coats.
   D. Epoxy primer should never be applied over a wash primer.

5. **Which statement is true about zinc chromate primer?**
   A. Green zinc chromate may be used on aluminum, but yellow zinc chromate may be used only on steel parts.
   B. Zinc chromate is a good corrosion inhibiter, but it does not bond to the metal as well as some other primers.
   C. Zinc chromate protects the metal by forming an absolutely watertight film over the surface.
   D. Zinc chromate is recommended as a primer for use under acrylic lacquers.

6. **Polyurethane finishes normally do not look good immediately after spraying. Why is this?**
   A. The surface soon dries to touch, but the film actually flows for several days.
   B. The surface was not sanded smooth enough before the polyurethane was sprayed on.
   C. The material was probably too thin.
   D. The material was sprayed on using too low atomizing air pressure.

7. **Who is nitrate dope preferred as a first coat, even though the subsequent coats will be butyrate?**
   A. Nitrate dope costs less than butyrate.
   B. Nitrate dope is more fire resistant than butyrate.
   C. Nitrate dope protects the fabric from mildew and fungus.
   D. Nitrate dope encapsulates the fibers of the fabric better than butyrate.

8. **What is the purpose of the aluminum dope in a fabric finishing system?**
   A. It protects the fabric and dope from the ultraviolet rays of the sun.
   B. It provides a good bond for the colored dope.
   C. It serves as a sanding coat to fill in all the rough spots in the finish.
   D. It is required so the airplane can be picked up on air traffic control radar.

9. **What is true about the use of polyurethane finishes on fabric?**
   A. The polyurethane should not be applied until the dope has cured for more than 48 hours.
   B. A polyurethane finish on the fabric does not require the use of any dope or other prime coats.
   C. When patching a polyurethane finished fabric, polyurethane enamel may be used as the adhesive to bond the patch to the surface.
   D. Polyurethane should never be used over a doped material.

10. **What will cause a spray gun to produce a crescent- or banana-shaped spray pattern?**
    A. Too much atomizing air pressure.
    B. A plugged hole in one of the wing ports.
    C. Not enough atomizing air pressure.
    D. A plugged atomizing air port.

11. **How much should acrylic lacquer be thinned?**
    A. Ten parts lacquer to approximately one part thinner.
    B. One part lacquer to approximately three parts thinner.
    C. Four parts lacquer to approximately one part thinner.
    D. Four parts lacquer to approximately five parts thinner.

12. **What is the best way to clean a spray gun in which dope has been allowed to harden?**
    A. Soak the entire gun in a pan of thinner.
    B. Throw the gun away; there is no practical way to get the dope out.
    C. Disassemble the gun, and soak all the metal parts in acetone or MEK.
    D. Heat the gun; the hardened dope will soften and run out.

13. **When building up a coat of sprayed-on material, how much should each pass overlap the previous one?**
    A. By all but two or three inches.
    B. By two or three inches.
    C. About half way.
    D. They should not overlap, just spray beside the last pass.

14. **How is fresh acrylic overspray prevented from causing a surface defect?**
    A. Sand it off before spraying on the next coat.
    B. Allow the overspray to dry, then spray on another coat of acrylic lacquer.
    C. The slow drying rate of acrylic allows the overspray to sink in the finish without causing any problems.
    D. Spray the overspray with a coat consisting of two parts thinner and one part retarder.

15. **What is true about the buildup of static electricity on an aircraft component while it is being sanded?**
    A. The component should be grounded so a buildup of static electricity cannot cause a spark.
    B. Static electricity will build up on a fabric-covered surface only, never on a metal surface.
    C. Rubber-soled shoes should be worn to prevent the buildup of a static charge on the body of the painter.
    D. Static electricity is more likely to build up while you wet-sand a surface, than when dry-sanding it.

16. **Color-matching metallic finishes may be quite difficult. Which statement is true about this operation?**
    A. Metallic finishes are usually sprayed directly over a green primed surface.
    B. Too heavy an application will produce a finish that is too bright.
    C. The use of too high an atomizing air pressure will make the finish too dark.
    D. If it is shot on too thin, with too much air pressure, it will be too light or too bright.

17. **Air-impingement haze in an acrylic finish may be caused by:**
    A. Too high an atomizing air pressure because of insufficient thinning of the material.
    B. Too much thinning of the material, and too low an atomizing air pressure.
    C. The use of an improper thinner.
    D. The use of too much thinner in the dope.

18. **What is the cause of dope blushing?**
    A. The use of too high an atomizing air pressure.
    B. Doping under conditions of too high humidity.
    C. The use of too much retarder in the dope.
    D. The use of too much thinner in the dope.

19. **What would likely cause flat black lacquer to dry glossy in spots?**
    A. Applied too thin.
    B. Applied too thick.
    C. Applied under conditions of too high humidity.
    D. The use of too high an atomizing air pressure.

20. **What determines the shape of the spray pattern of a gun used with a pressure pot?**
    A. The amount of air pressure used.
    B. The amount of air allowed to flow through the wing ports.
    C. The amount of pressure on the pressure pot.
    D. The amount the gun trigger is pulled.

ANSWERS TO FINAL EXAMINATION
Aircraft Painting And Finishing

1. C
2. C
3. B
4. C
5. B
6. A
7. D
8. A
9. A
10. B

11. D
12. C
13. A
14. D
15. A
16. D
17. A
18. B
19. B
20. B